W9-BHL-289

How Data Happened

How Data Happened

A History from
the Age of Reason to the
Age of Algorithms

Chris Wiggins and Matthew L. Jones

W. W. NORTON & COMPANY

Celebrating a Century of Independent Publishing

For information about permission to reproduce selections
from this book, write to Permissions, W. W. Norton & Company, Inc.,
500 Fifth Avenue, New York, NY 10110

For information about special discounts for bulk purchases,
please contact W. W. Norton Special Sales at
specialsales@wwnorton.com or 800-233-4830

Manufacturing by Lake Book Manufacturing
Book design by Lisa Buckley
Production manager: Anna Oler

ISBN 978-1-324-00673-2

W. W. Norton & Company, Inc., 500 Fifth Avenue, New York, N.Y. 10110
www.wwnorton.com

W. W. Norton & Company Ltd., 15 Carlisle Street, London W1D 3BS

1 2 3 4 5 6 7 8 9 0

To our families, who made this book possible

CONTENTS

PROLOGUE   ix

PART I

CHAPTER 1   The Stakes   3
CHAPTER 2   Social Physics and *l'homme moyen*   18
CHAPTER 3   The Statistics of the Deviant   35
CHAPTER 4   Data, Intelligence, and Policy   54
CHAPTER 5   Data's Mathematical Baptism   78

PART II

CHAPTER 6   Data at War   101
CHAPTER 7   Intelligence without Data   120
CHAPTER 8   Volume, Variety, and Velocity   141
CHAPTER 9   Machines, Learning   175
CHAPTER 10   The Science of Data   196

PART III

CHAPTER 11   The Battle for Data Ethics   233
CHAPTER 12   Persuasion, Ads, and Venture Capital   257
CHAPTER 13   Solutions beyond Solutionism   284

ACKNOWLEDGMENTS   309
NOTES   313
INDEX   353

PROLOGUE

One April morning in 2018, as the spring sunlight streamed into the eastern window of a seminar room in Schermerhorn Hall at Columbia University, I (Wiggins) went to the chalkboard to explain quantitative reification, the magical process whereby a numerical correspondence to empirical observation becomes a thing. Armed with the story of Adolphe Quetelet, who aimed to reveal the ideal man using data he obtained about the physical measurements of Scottish soldiers, I traced on the blackboard the immortal "normal curve." Known to mathematicians as the Gaussian curve, and contested as the notorious "bell curve" of IQ tests, the normal curve signifies to natural scientists that data has revealed something real, even something transcendent. I turned to the students, hoping to see in their eyes that they shared my excitement. One met my gaze and, turning his palms to the heavens, asked: "Can we talk about Facebook now?"

That morning, newspapers and digital news feeds alike heralded a hot fire about to burn in Washington, one that would melt down all concealment. The irreverent CEO of a culture-changing tech company in Silicon Valley was being called before the United States Senate. On behalf of all citizens, the senators sought to understand how the personal data of millions of people, including students such as ours, was compromised, *The New York Times* explained, used for ill ends that violated our norms about privacy and our political process.[1] By the end of the week's congressional testimony, students recognized the extent of the gap between how our elected officials understood their digitally mediated reality and their personal knowledge from growing up with algorithms.

The story of data is replete with contests: contests to define what is true, contests to use data to advance one's power, and, on occasion, contests to use algorithms and data to shine a light into darkness and to empower the defenseless. This book grew out of our teaching hundreds of inquisitive students, along with our own experiences, as a historian of science and as a practicing data scientist, and as citizens trying to understand how we came to live in this algorithmically mediated reality and how we might choose to live differently. Like all users, developers, and subjects of technology, we are trying to make sense of where it is all headed as well as how we collectively will shape that future. We've attempted to tell a story of ideas and technologies but also a history of truth and power.

Putting down the chalk, we agreed that Quetelet would have his day. But first, we would need to explain how an obscure Belgian astronomer fits in with the story of data: how data and the means for analyzing it would move from a concern of the state to universities, the military, and private corporations.

We use "data" here as shorthand for the expanse of data-driven algorithmic decision-making systems surrounding us nearly everywhere. We explore how data was created and curated as well as how new mathematical and computational techniques developed to contend with that data serve to shape people, ideas, society, military operations, and economies. Along with data comes power, including the power to shape what is perceived to be true. Although technology and mathematics are at its heart, the story of data ultimately concerns an unstable game among states, corporations, and people.

And so, on that morning, we spoke not just of data, but of the stakes for a world mediated by data.

Background

The idea for a class on how data happened was born in November 2015, at a small dinner conversation with a few Columbia undergraduates drawn from a mix of engineering and humanities backgrounds. At the time, we conjectured students were very interested in the history of data science, and we thought that our combined, complementary perspectives would give a useful view, with material new both to the engineers and to the non-technologists alike. When we taught the class for the first time in January 2017, we quickly realized that the students were not only interested in how we got here but were searching for an analytic and actionable framework for understanding the ethics and politics of data.[2]

By "politics" we don't mean a narrow definition as "voting" but "of or relating to the dynamics of power." Our goal is to provide a framework for understanding the persistent role of data in rearranging power: corporate power, state power, and people power. Our historical arc offers critical leverage that points us toward a shared understanding of the present as well as the weapons and tools at our disposal for shaping the future.

This Book

Every history must begin somewhere, and we found a useful starting point to be the end of the eighteenth century, around the time the word "statistics" first entered the English language. Our story tacks between the hard work of collecting data, including building infrastructure to collect and make it public, and the development of new mathematical and computation techniques for studying data—new ways of understanding and making claims about that data and using those claims to make decisions, often profoundly changing lives, for better and for worse. In each chapter we consider one intellectual transition. We discuss how a new technical and scientific capability was developed; who supported, advanced, or funded this capability or transition; how this transition was contested; and how this new capability rearranged power—changing who could do what, from what, and to whom.[3] We focus not only on rearrangements of military or financial power, but more generally on those transitions with an ethical and political valence: those in which data affects rights, rearranges harms, or supports—or thwarts—justice.

How Data Happened begins with data in the service of statecraft, before turning to the usage of data to improve society and the mathematical baptism of data with the creation of a new academic field called "mathematical statistics." The second part opens with the martial application of data in World War II for codebreaking, coinciding with the birth of digital computation. We follow the thread from Bletchley Park in England to Bell Labs in the United States, and to the business and engineering applications of data in the wake of World War II. Transitioning from corporate power to the reactions in state power and in "people power," we explore the impact of digital, personal record keeping on our understanding of privacy, particularly the public desire for privacy as a defense from overreaching state power in the 1970s. We trace the first birth and death of the field of "artificial intelligence" and its renaissance, rising from the ashes in the form of "machine learning" based on ever growing repositories of data about citizens, consumers, and military adversaries alike.

The last section of the book connects this past to our present and future. We discuss how data and power moved from a state concern to a corporate concern, by looking to the financial arrangements and business models that have allowed single corporations to dominate entire sectors rapidly with the help of data-empowered technologies. A contested debate over ethics has framed many of the potential remedies to corporate power; we trace the history of applied ethics in research and how it has impacted the way that data-empowered algorithms are deployed as products, shaping our personal and political reality.

Finally, we discuss the future. However difficult it may be to make predictions, one incisive way to organize our understanding of the future is to describe the present contests among powers along with the arenas in which these contests will be decided. We close the book by looking at what we consider to be the most important present contests among corporate power, state power, and people power, along with the possibility of new forms of solidarity. The resolutions of these contests will shape our collective future, leaning more toward justice—or perhaps not.

Our goal here is an actionable understanding of history. We will not shy away from our own roles as citizens, technologists, and individuals; we are users of these products and— as noted as early as the 1970s, since we are in an advertising economy, therefore we are also the product.

We bring to the book two complementary perspectives, each with limitations and biases. Wiggins has been developing machine learning methods for understanding biology and health for over twenty years as faculty at Columbia and, since 2013, developing and deploying machine learning methods and products as chief data scientist at *The New York Times*. On the other side of C. P. Snow's "two cultures," Jones is a historian of science who has tracked how mathematical ways of thinking and arguing became a crucial authoritative way to study nature and politics from the "Scientific Revolution" of the seventeenth century forward. Particularly when examining how many uses of data amplify disparity, we draw heavily upon the illuminating writing of the many scholars and activists who have exposed these processes. Many, if not most, of the most trenchant and luminous critics crucially come from backgrounds and have experiences quite different from ours—two tenured white male academics. Our work builds upon and celebrates their labor and insight throughout. We will point to excellent literature on the global impact of data-empowered algorithms and technologies—and to the histories of data in the organization of our societies, economies, and educational institutions. Our more contemporary material focuses primarily on the United States. We have provided endnotes to reflect not only where to find out more about the topics we cover in class and write about in our scholarly publications, but many important works, including the scholarly literature, that we encourage readers to engage for deeper understanding.

We seek to give a clear picture of historical as well as current tensions among corporate power, state power, and people power, focusing on the role of data in establishing truth and shaping the contests among these powers. We hope to show how we collectively got here, to illustrate the small coincidences, subjective design choices, and deceptions that ossified into what only *seems* like things that "must be that way." Understanding these transitions and contingencies will reveal how similar problems were solved in the past. This will, in turn, help us picture how we could break and reset the bones of systems that sometimes empower the defenseless—yet have more often strengthened the empowered.

By showing how apparently immutable results hinged on past choices, we can see how we can collectively choose a different future.

PART I

CHAPTER 1

The Stakes

Technology is neither good nor bad; nor is it neutral.
                          —Kranzberg's first law of technology, 1986

I teach a course at the University of Michigan called "the Internet
is a trash fire," and I don't have to explain to anybody what that
means. . . . We put up with this for a long time; we don't seem to
know anything different.
                          —Lisa Nakamura, 2019

I n December 2014 at the Palais des Congrès de Montréal,
the computer scientist Hanna Wallach advocated for a revolution before an audience of technologists, lawyers, and
activists. Speaking to top computer scientists working on
"machine learning," she proposed that her own field desper-
ately needed to interrogate how the algorithms they were
developing, and the technologies the algorithms empowered,
challenged our values of "fairness, accountability, and transparency." While philosophers, sociologists, and lawyers had
been sounding the alarm for years, here, an anointed member of the technical community, with a coveted research position at Microsoft, drew on that critical work and called for
colleagues to improve their research, to do better work precisely by recognizing the need for fairness and accountability from their algorithmic systems.

    Far from a protest in the wilderness, Wallach's talk, delivered at the most important conference in applied machine learning, was a posting of theses directly on the door of the cathedral. Wallach diagnosed the problem—one outside the traditional disciplinary scope of computer science. Admitting that the solutions to the problem would not come from within computer science, she instead demanded collaboration with those from other fields. "Few computer scientists or engineers," Wallach explained, "would consider developing models or tools for analyzing astronomy data without involving astronomers. So, why, then, are so many methods for analyzing social data developed without the involvement [of] social scientists?"[1]

Wallach urged a deeper recognition of the ways that biases creep into models created by machine learners and warned of the risks inherent in studying data sets simply because they were available. As an example: while it's relatively straightforward to obtain and analyze data from Twitter users, these data are hardly representative of, say, the US population overall. She urged researchers "to start thinking outside the algorithmic boxes typically embraced by the machine learning community and instead focus on the opportunities, challenges, and implications involved [in] developing and using machine learning methods to analyze real-world data about society."[2]

Analyzing real-world data about society was, by the time Wallach was speaking, already the heart of the business models of the internet giants Google, Facebook, and Amazon—not to mention central to the intelligence agencies of the United States, the United Kingdom, Israel, and China. Suffice to say, these corporations and agencies rarely incorporated these questions of fairness and accountability animating Wallach's talk. The issues at heart were not simply academic nor merely a question of a research community shifting its focus.

Swimming against the tide of internet utopianism after 2000, a lively group of social scientists had signaled concerns about the data-driven internet in commerce, education, and governance. At a 2011 symposium held at Oxford, danah boyd and Kate Crawford argued, "The era of Big Data has begun." In front of an audience including such luminaries as Vint Cerf, one of the "inventors of the internet," the researchers sought to provoke the community to think more critically about the incipient age of big data:

> Will large-scale search data help us create better tools, services, and public goods? Or will it usher in a new wave of privacy incursions and invasive marketing? Will data analytics help us understand online communities and political movements? Or will analytics be used to track protesters and suppress speech? Will large quantities of data transform how we study human communication and culture, or narrow the palette of research options and alter what "research" means?[3]

Drawing upon earlier critical voices like Oscar Gandy Jr., Wendy Chun, and Helen Nissenbaum, researchers began documenting the real-world effects of corporations and governments failing to face these troubling questions—and called for dramatic change.[4] Without claiming to capture adequately this enormous body of work, let us mention a few key examples.

In 2013, Safiya Noble, then a professor at the University of Illinois at Urbana-Champaign, now a MacArthur "Genius" fellowship awardee, published an excoriating look at Google search. "Commercial search implodes," Noble wrote, "when it comes to providing reliable, credible, and historically contextualized information about women and people of color, especially Black women and girls." Ostensibly bias-free technology quickly reproduced—and reinforced—racist and sexist biases toward Black women. "Continued study of these phenomena," she wrote, "is an opportunity to contest the alleged neutrality of technology, while creating new opportunities for social justice and fair representation online."[5] In 2016, mathematician Cathy O'Neil described her own journey from academic to Wall Street employee to critic of data and algorithms gone unchecked. In her *Weapons of Math Destruction*, she explored how the incentives in data science undermined the humanity of its subjects: "The inclination is to replace people with data trails, turning them into more effective shoppers, voters, or workers to optimize some objective. This is easy to do, and to justify, when success comes back as an anonymous score and when the people affected remain every bit as abstract as the numbers dancing across the screen."[6] Without changing these incentives, data science, for all its promise, would dramatically alter the goals of organization after organization, from universities, to medicine, to social welfare, at the expense, most of all, of the least powerful members of society. Around the same time as these critical diagnoses, the revelations of the vast expansion of the spying apparatus of the United States and its "Five-Eyes" allies around 9/11 by Edward Snowden reanimated longstanding concerns about mass government surveillance, previously raised by earlier generations of whistleblowers, journalists, and civil libertarians. The American NSA (National Security Agency) and the British GCHQ (Government Communications Headquarters) sponsored and benefited from the dramatic explosion of academic research and commercial developments on the collection and analysis of data. Earlier ways of understanding the dangers of surveillance demanded a more critical legal and technological analysis of the violations of privacy that ever-greater amounts of data and sophistical analytical techniques made possible. A tremendous surge of critical diagnoses of the new centrality of data and algorithms emerged alongside these prominent interventions.

Within a few years of the surge of critical concern, the likes of Google, Facebook and IBM all had in-house AI ethicists. These firms—economic powerhouses financially equivalent to many nation-states—had quickly appropriated the critical movement, hiring many of the most brilliant critics, but often silencing or disempowering them when the criticism got too hot. It was as though the Pope had hired Martin Luther and set him up in a corner office of the Vatican, while the indulgences spurring the Reformation went on relatively unabated. The researchers hired by the likes of Google, such as Timnit Gebru and Margaret Mitchell, too often found themselves challenged as they struggled not to be co-opted. The computer science research community, for its part, turned these deep concerns about fairness into new algorithmic puzzles, but too often carefully gated them away from reflections upon power.

Scholars like Wallach, Noble, and O'Neil among many others all keenly saw how new algorithmic systems easily reproduced in their automated judgments the systemic inequalities of yore at an unprecedented rapidity and scale. New capabilities yielded new powers. These powers threatened to entrench many of the inequalities that so many societies had long struggled, with checkered and uneven success, to undo. Too often these new technologies reinforce existing forms of structural inequality and difference, in what Princeton professor Ruha Benjamin describes as "a set of technologies that generate patterns of social relations" that themselves "become black boxed as natural, inevitable, *automatic*."[7] The veneer of objectivity that comes from technologies using data serve, Ruha argues, to "encode" inequity.

And this inequality extends into the firms building algorithmic systems: the capacity to deploy these technologies at vast scale today is only truly available to the best-resourced corporations and governments, as ex-Googler Meredith Whittaker has stressed.[8]

> Science fiction writer William Gibson is believed to have said, "The future is already here—it's just not very evenly distributed." I believe that too, but in a way opposite to what I think he intended. People who live in low-rights environments—poor and working-class communities, migrant communities, communities of color, religious or sexual minorities—are already living in the digital future, especially when it comes to high-tech surveillance and discipline.
>
> —*Virginia Eubanks*[9]

The warnings of Wallach, boyd, O'Neil, Noble, and other scholars and activists were neither the first nor the only examples of what is now widely recognized as illuminating the downsides of the data deluge in which we live. As Lisa Nakamura notes, "I teach a course at the University of Michigan called 'the Internet is a trash fire,' and I don't have to explain to anybody what that means."[10]

Early generations of scholars and legal activists from the 1960s onward had signaled the dangers of the accumulation and automatic analysis of data to privacy and noted how such inquiry often exacerbates existing inequalities. Sociologists had raised concerns about the impact of data-empowered algorithms on democratic politics. Zeynep Tufekci warned in 2014 of the rising "capacity of those with resources and access to use these tools to carry out highly effective, opaque and unaccountable campaigns of persuasion and social engineering in political, civic and commercial spheres."[11] While political persuasion and quantitative or "performance" marketing were not new in 2014, their combination with political influence operations and "microtargeting"—the ability to optimize the delivery of different digital messages to individuals—opened up the possibility of fracturing the realities of the electorate into what Renée DiResta later called "bespoke realities."[12] The US presidential election of 2016 brought home the realities of such microtargeting, especially around the company Cambridge Analytica and its exploitation of data gleaned via Facebook. Public imagination and fears were well captured by the story of what appeared to be a harmless personality quiz on Facebook being leveraged to "engineer the public," as Tufekci put it, leading to congressional testimony by Facebook's CEO Mark Zuckerberg in the spring of 2018, reassuring lawmakers that, despite the fears of their constituents, their data was not for sale and all was well in Menlo Park. (The lawmakers' responses were not always reassuring to the electorate, however: one confused senator asked if Facebook was "the same as" Twitter, and another senator asked Zuckerberg to clarify if the firm sells ads—their primary business.)

Fears of the power of algorithms to drive "computational politics" at the scale of entire nation-states mirrored concerns raised about the power of algorithms to create what Virginia Eubanks calls a "digital poorhouse." Her 2018 book *Automating Inequality* traces three stories about algorithms disempowering the poor and the needy. She shares how opaque algorithms used by the state—some complex, some simple—serve to exacerbate social inequality, delivering harms at scale to those least empowered to defend themselves and critique their use. Warning us of the way unjust harms can be used on all but the most empowered, one of her informants chides her: "You should pay attention to what happens to us. You're next."[13] Tales of predictive policing have moved from dystopian science fiction to reality in trials in the United States and beyond, perhaps most notoriously Chicago's "Strategic Subjects List."[14] When data-empowered algorithms are deployed by police forces, they have already led to wrongful arrests and incarceration, with growing concern about biases in their design and deployment. Health workers and public health officials alike face harassment, denial, and counternarratives. Nation-states face destabilization and disinformation at enormous scale.

The potential threats from algorithmic decision systems replete with granular data on increasingly large numbers of people are in many ways new. They enable governments and corporations to know about our everyday activity at an entirely new scale: techniques previously directed at small groups, often the most marginalized or dissident, can be applied to the entire population. They constitute an unprecedented intimacy in that they power our interpersonal communications, our sources of news and information, and even algorithmically moderate our relationships.[15] This makes such systems (including algorithms recommending movies and entertainment, news, or romantic partners) all the more potentially damaging in the cases of either abuse or poor design. In the case of disinformation, for example, the nature of open information platforms means that the dangers of data-empowered algorithms come from not only nation-states but our neighbors.

Within academia, responses to the explosion of machine learning systems are widely varied, between enthusiasm and alarm, with growing participation from varied technologists, social scientists, and humanists. Yet the tightening relationship between industry and academia rightly raises hackles: as the size of sponsored industrial research grows, rivaling that of traditional government funding, a type of "capture" is enabled in which research critical of technology companies is thwarted actively or merely disincentivized by the fear of lost financial support.[16]

Outside this industrial-academic complex, among activists and among departments and faculty not directly or financially benefiting from data-empowered technology companies, there is not just concern but open activism to limit the power of technology companies, either by advocating for state regulation or in the form of "private ordering," including discouraging others from working for or with technology companies that harm society.[17] Corporations have responded, particularly over the past few years, in a variety of ways, some old and some new. Traditional responses, such as government lobbying and public relations campaigns to the public, have grown in intensity and financial scale—and even deviousness. Responses particular to concerns over algorithms include both technical fixes under the label "fairness" as well as establishing "AI ethics" teams, roles, or principals. Both have drawn mixed responses from the US Congress, from the concerned, and from the critical. Neither has yet succeeded in dramatically changing internal processes at the most powerful of companies. Law professor Frank Pasquale has warned for many years that enormous corporations co-opt genuinely important values such as transparency. In 2016, a coalition of researchers produced important "Principles for Accountable Algorithms" that argued,

> Automated decision making algorithms are now used throughout industry and government, underpinning many processes from dynamic pricing to employment practices to criminal sentencing. . . . Accountability in this context includes an obligation to report, explain, or justify algorithmic decision-making as well as mitigate any negative social impacts or potential harms.[18]

Far more is necessary, as most of the writers we invoke would probably now agree; we must have robust institutional forms that enable a holding to account, not simply force the production of an account. Tackling the dangers and promise of algorithmic systems demands concentrated political action, capable of affecting who data empowers and who it does not; and it demands a clear understanding of how contingent—how not set in stone—our current state of affairs is. The better we understand the genesis of those systems, the better equipped we collectively will be to contest, defy, and put them to more just uses.[19]

History and Critique

We opened the chapter with a law of technology, Kranzberg's first law, named after a historian of technology who wrote in 1986:

> my first law—Technology is neither good nor bad; nor is it neutral—should constantly remind us that it is the historian's duty to compare short-term versus long-term results, the utopian hopes versus the spotted actuality, the what-might-have-been against what actually happened, and the trade-offs among various "goods" and possible "bads." All of this can be done only by seeing how technology interacts in different ways with different values and institutions, indeed, with the entire sociocultural milieu.[20]

The growing voices of alarm around data today compete with voices of optimism, fired by clear technological advances in using data to make sense of our world and to facilitate everyday activities. As we write, we accept as normal a level of speech-to-text and automated spell-checking that far exceeds the best efforts of even a couple of years ago. In research, we've seen marked advances in the ability to predict protein folding, or to identify disease from clinical image data and genomic data. Promises of great futures with self-driving cars and personalized or "precision" medicine fill the tech press and the marketing materials of tech companies alike. And we need not accept the overblown marketing hype around data to recognize the profound effects, many unintentional, wrought by these technologies embedded within our social, political, and economic systems.

With these developments come challenges to many authorities and many professions, from scientists to advertisers, physicians to lawyers. The story of machines replacing hand workers at a vast scale is as old as the Industrial Revolution; and now the machines are coming for elite white-collar workers—and quickly. Doctors, for example, are increasingly recognizing that machines will soon do more central diagnostic practices: "Predicting where disruption is most likely to occur is hard, but if it feels routine today, then it is likely to be a target for the machine tomorrow. Clinical specialties like radiology might not disappear, but they certainly will be heavily transformed."[21] Even as these traditional professions are challenged, a global range of new workers will be required to enable the systems doing the work.

In the last ten years, we've seen increasing threats to individuals' rights, harms, and justice from corporations and governments alike; at the same time, we've also witnessed great developments, in our personal lives, and in research, and the promise of technological benefits to come. It is also, by now, clear that those in power—particularly state and corporate power—will not be giving up on data-empowered capabilities without intense pressure and advocacy. We must take on Kranzberg's challenge to understand the short- and long-term results, as well as to understand the ways our choices constitute small and large reorderings of power.

History as Solvent

Powerful forces often are reticent to investigate the historical genesis that made them possible—or even dominant. Complex histories unsettle the obviousness, the legitimacy, of their power.[22] In looking at the far from obvious ways technologies come to prominence, history unsettles the idea that the growth of certain technologies themselves drives history, a view called "technological determinism." It has been very lucrative for many interested actors, for example, to claim that older views of privacy are outdated in the age of the internet, even that the internet itself causes the decline of privacy. Neither claim is true. But such stories offer a potent version of history, ubiquitous in debates around the internet, that legitimates the current order of things as necessarily so.

History can collapse into nostalgia about a more humane, better past, but it need not. Whatever the novelty, dangers, and scale of contemporary algorithmic decision-making, the emergence of soulless bureaucracies using quantitative measures has an often-dim history. Scholars from Michel Foucault and Bernard Cohn to Jacqueline Wernimont, Martha Hodes, Simone Browne, and Khalil Gibran Muhammad show how the quantification of peoples has a long history from the early nineteenth century onward in ways of ranking and classifying students, races, colonized peoples, enslaved people, soldiers, the poor, the mentally ill, and the incarcerated.[23] Historians like Sarah Igo, Emmanuel Didier, Dan Bouk, and Emily Merchant explore how surveys and censuses don't simply record: they constitute publics and populations; they enable forms of solidarity and types of governmental action—and inaction. Data is made, not found, and the process of procuring and analyzing it often dramatically loops back to shape the people under official scrutiny.[24]

Long before the SAT reduced college applicants to a score, the psychologist Charles Spearman proposed a mathematical "general intelligence" score to reduce intellect to a number; long before writers and Amazon could know how many online readers were engaging with their writing, the nineteenth-century mechanical engineer Frederick Taylor introduced scientific management to quantify worker output; many decades before both, sophisticated means of accounting and bookkeeping were at the center of plantation slavery, as historian Caitlin Rosenthal has shown.[25] And yet our societies benefit from the knowledge that rigorous quantitative studies can bring in organizing our social, medical, and political lives. Our trust in vaccines—to take one example—rests crucially upon a standardized quantitative process of assessing efficacy and gauging harm. Quantified measures can—and have—and do—provide accountability, but they've been turned against us with a feverish pace.

Key methods of numerical accountability emerged in large part as tools for resisting expert judgment ensconced in critical governmental, educational, and corporate entities. The science historian Theodore Porter argues that standardized forms of numerical accounting, such as cost-benefit analysis, arose to contest the power of *human* "black boxes," experts with authority grounded in traditional status and opaque forms of judgment. Numerical accountability, with its promises of transparency and rule-bound objectivity, typically gained prominence in conditions of distrust about experts.[26] The 1933 "Truth in Securities" act, for example, sought to bolster trust in capital markets through uniform standards of accounting and reporting, which were bitterly resisted by Wall Street and its accountants. Forcing banks years later to reveal their protocols for lending money revealed systematically racist criteria and forced creation of new criteria. Whatever its limits, making information about major institutions available using quantitative measures has been, since the nineteenth century, a formidable tool to check state and corporate power, particularly the opacity of expert decision-making in organizations. Such knowledge can enable people to challenge experts, to ensure fairness, to make decisions transparent.

And yet the mechanisms so useful for making powerful institutions more accessible to the public have long since been focused *on* the public. Rather than making powerful institutions transparent to us, algorithmic systems increasingly lay us bare to powerful institutions.[27] In the last four decades, corporations, universities, and governments have imposed numerical measures of accountability on individual employees and citizens at an ever-accelerating pace. Nothing is inevitable or trivial about imposing such systems of measure.[28] The last forty years have seen an explosion of techniques for insisting on numerical accountability adapted and applied in concrete contexts. Thanks to decades of effort, numerical measurement now saturates workplaces, from factories to universities, from Uber drivers to fisheries, through systems of metrics that make employees knowable to employers—and accountable to their bosses for nearly all their activities. Algorithms—often secret, proprietary, and rarely transparent—process this data to rank and to classify, to promote and to fire, to reward and to punish. Systems of accountability tend to impose themselves on lower rungs of institutional systems, falling predictably along socioeconomic and racial lines. Rather than rendering powerful institutions transparent to outsiders, these measures instead often render everyday employees and citizens transparent to powerful institutions. And they do so through algorithmic decision-making and classification usually removed from scrutiny.

We'll start, long before Google search and Uber rideshares, with the dreams of the Belgian astronomer who invented the body mass index and "social physics." His goals, we'll see, sound contemporary: to use the latest technologies of the day and collected data about individuals. And to empower those using these technologies to improve society itself.

CHAPTER 2

Social Physics and *l'homme moyen*

A nineteenth-century Belgian astronomer, Adolphe Quetelet, inspired Florence Nightingale. A prophet with a new vision for making law. An evidence-based vision. Write down, this Belgian advised, "what you expect from such and such legislation. After x years see where it has given you what you expected and where it has failed." Writing in 1891, Nightingale complained that lawmaking of her times involved no such data: "You change your laws and your administering of them so fast and without inquiry after results past and present, that it is all experiment, see-saw, doctrinaire, a shuttlecock between two battledores."[1] This Belgian, the "founder of the most important science in the whole world," Nightingale explained, had provided "the one science essential to all political and social administration." Quetelet "did not live to see it perceptibly influence, in any practical manner, statesmanship—of which there is none without it—or government."[2] While this new science—these new capabilities—had not yet rearranged power, Nightingale was sure that they should. Our world involves just such a rearrangement of power.

Quetelet gave us the body mass index, the idea of the statistically average person, and, above all, dramatically altered how we think about societies. Quetelet "was fond of numbers," the philosopher Ian Hacking quipped, "and happy to jump to conclusions."[3]

For all his radical new ideas for thinking about society, Quetelet wanted to avoid radical disruption. His time had seen far too much of that, from the French and Haitian revolutions to the Napoleonic Empire. In 1830, revolutionaries occupied his new astronomical observatory in Brussels, to his great dismay. "Our observatory," Quetelet wrote a friend, "has just been converted into a fortress."[4] Violent revolution had lost its appeal. Obsessed with applying mathematics to society, Quetelet sought to create a new science of nonrevolutionary change. In political as well as social life, he explained, "abrupt movement" causes a wasteful loss of force. "This principle is advantageous to the partisans of a revolution," he noted. Society needed reform. Revolution wasn't the way. "Abrupt movements are never made without a certain loss of live force. This principle is not advantageous to the partisans of a revolution, unless they impel forces in a more useful direction [and] consent to lose a portion [of these forces]."[5]

A new scientific politics, based on data about people, should rearrange power. Gradually. Without occupying buildings. Without disruption.

Bureaucracy, budget problems, and construction challenges long delayed the completion of his observatory. While he waited to survey the sky, Quetelet drew on the best techniques for studying the stars to think through observations about people.[6] Directly influenced by the triumphant late eighteenth-century successes of physical and astronomical models of heaven and earth, and dismayed by the early nineteenth-century political and martial upheavals of power in Europe, Quetelet sought to create a new "social physics."

But numbers were not the obvious way to understand humanity or power relations in 1830—it was no accident that Nightingale was calling for social physics many decades later.

Vulgar Statistics

How did we come to think that numbers are essential to understand the world and the lives of its peoples? From artists to anthropologists, from novelists to grand viziers, critics have long said no to quantification. "These stupid fellows," a German polemicist wrote in 1806, "disseminate the insane idea that one can understand the power of a state if one just knows its size, its population, its national income, and the number of dumb beasts grazing around."[7] Real statistics, genuine knowledge of the state, he maintained, unlike its "vulgar" cousin, involved careful description and knowledge of history. Such investigation transcended the material to grasp the moral and spiritual texture of different countries. Tabulating mortality for moral guidance, news, and profit had grown since the late seventeenth century, but artlessly applying such crude tables to major questions of statecraft was anathema. Number crunchers were "table-statisticians," not real statisticians. Numerical depiction "does not touch upon the spiritual forces and relationships of states, morals, the divine." Such statisticians "see quality not at all, but only quantity."[8]

Two hundred years later, former Republican speechwriter and *Wall Street Journal* op-ed contributor Peggy Noonan similarly decried a state of affairs she found most ridiculous:

The other day a Republican political veteran forwarded me a hiring notice from the Obama 2012 campaign. It read like politics as done by Martians.

The "Analytics Department" is looking for "predictive Modeling/Data Mining" specialists to join the campaign's "multi-disciplinary team of statisticians," which will use "predictive modeling" to anticipate the behavior of the electorate.[9]

By 2016 both parties had formidable data operations in this vein.

Numbers haven't always been the obvious way to understand and to exercise power. How did it get that way? Why do we now turn to them? And once computerized, how are they pathological, as well as liberating? How did the mathematical analysis of data about people and things come to be such a dominant way to understand and to control the world, to predict and to prescribe? The critics of numerical statistics at the end of the Enlightenment well understood that data is profoundly artificial. As Lisa Gitelman noted some years ago, "raw data is an oxymoron," as all data collection comes through human choice about what to collect, how to classify, who to include and to exclude; all collection involves cognitive biases and radically different infrastructures for categorizing, storing, and processing that information.[10] Data is made not found, whether in 1600 or 1780 or 2022.[11] How did such data become powerful? How did the structures to collect, to store, and to analyze it get built? How did arguments using it become so convincing—and even legally necessary?

In Europe in the eighteenth century, war, taxes, and sometimes life and usually death dominated the concerns of rulers. Eighteenth-century Europe saw continual bloodshed with punctuations of peace, often extending to brutal conflict in the Americas and elsewhere. War required money; money required taxes; taxes required growing bureaucracies; and these bureaucracies needed data. The burgeoning states of Enlightenment Europe needed to know what resources they had: people, land, precious metals, and industries. *Statistics* was originally knowledge of the *state* and its resources, without any particularly quantitative bent or aspirations at insights, predictive or otherwise. From 1780, an explosion of counting took off, which Ian Hacking memorably described as "an avalanche of numbers."[12]

This new, highly numerical statistics threatened the older ways of understanding rule and understanding people. Rather than basing the organization of the state upon the classics of political philosophy and using the history of states ancient and modern as guides, advocates of the new statistics focused on the quantitative descriptions of "land and people" relevant to guiding the ruler. Reforming officials, armed with new ways of studying the people and the state, tried to convince rulers that they and their methods were necessary for the growth and health of the state. They sought to describe, and to interpret, these descriptions as providing suggestions for policy. Enumeration was never "neutral" but designed with goals in mind, and interpreted in ways that suggested policy, particularly allocation of resources. Near the close of the eighteenth century, the new United States of America enshrined the census in its most fundamental law, the Constitution. Then, as now, numbers were political.

The history of personal data—its collection and interpretation—often involves the powerful reinforcing of political, military, colonial, and industrial power. Given long traditions of collecting information about lands and peoples in China, in Incan space and elsewhere, such practices were not exclusive to late Enlightenment European states. Yet quantification gained a radical new centrality in Europe and then the United States and colonies worldwide from the eighteenth to the twentieth century.[13]

Statistics was initially a new technology for states at a moment of increasing industrial, commercial, and martial competition. Heirs of Malthus, we worry about overpopulation. European thinkers of the eighteenth century, on the contrary, were anxious about underpopulation, often attributing economic underdevelopment to it. Monarchs and their advisors came to view the strength of states—and of "races"—as quantified by the size and vigor of its population.

Regularly published bills documenting the causes of deaths in parishes comprised some of the earliest recognizable collections of numerical data in seventeenth-century England. In recasting death into numbers, these bills, Jacqueline Wernimont explains, "produced an ironically idealized world in which the reporting of epidemic disease and mass death appeared as clean and orderly as an account book."[14] From the eighteenth century onward, Europeans dramatically began to record abundant data and create new mathematical tools to examine these data to strengthen governments, influence policy, and persuade their peoples. As the accumulation of numbers accelerated, more and more facets of human lives were recorded in abstract numerical terms. From the start, governments, churches, and private statisticians tabulated numbers about deviance, death, crime, and sickness. Institutions new and old recorded details about the course of life and death, and—then as now—people running afoul of the law left traces. Statistical thinking from the 1700s onward rested fundamentally on the explosion of the collection of data about states, their people, and, quite often, people deemed to be deviant.

This collecting of numbers was initially largely a descriptive affair, with scant calculation or mathematical work along the way. When the Statistical Society of London was founded in 1834, they chose a seal with the words *aliis exterendum* ("to be threshed out by others"). They sought "simply to gather the facts, leaving it to others" to interpret them.[15] Others, however, worked to develop the new tools for making sense of all these numbers and making arguments based on them.[16] Financiers, scientists, and bureaucrats alike began, from the eighteenth century onward, to develop new mathematical and visual means for making sense of this data and for making claims based on it, whether to convince investors to pony up or to affect policy. While our Belgian astronomer Quetelet may have had the greatest impact on subsequent statistics, German statesmen called "cameralists," English demographers and financiers, and others devised ways to rework statecraft and economies using new forms of data-driven analysis.[17] Having moved dramatically away from its qualitative roots, the term "statistics" came to incorporate, on the one hand, the accumulation of data, primarily numerical data, about everything from people to climate, and on the other hand, a set of powerful, beguiling, and often misused mathematical tools to draw conclusions and analyze data.[18]

Just as data-driven analysis "disrupted" the marketing of goods at your local grocery in the 1990s, empirical analysis of the population, production, and acres under tillage challenged older ways of knowing in order to rule. The study of data threatened to displace other forms of expertise, from science to shop floors to the drug store. Rather than lush descriptions of countryside, a counting of flora and fauna. Rather than an ethical discussion of values, an attempt to model the effects of a given policy quantitatively. Rather than the gruesome reality of death, tables of mortality statistics. Rather than expertise about potential desires of consumers, the collection and analysis of every purchase. Rather than the clinical experience of individual physicians with a drug, randomized trials to gauge effectiveness and safety. Rather than a judgment of the character of a student applying to college, the use of standardized tests to supply an "objective" measurement.

The birth of statistics in the modern sense comes from

the realization that fusing data and mathematical analysis could serve power—but also, at times, could check power.

Enter our Belgian astronomer, Quetelet.

Astronomer Looking at the Social World

Knowing *states* or *countries* quantitatively soon thereafter led scientists to try to know *human beings* quite differently, to change the way we understand ourselves—as moral people, as physical beings, as social beings. Governments and other institutions began collecting data on death, crime, and suicide in fits and starts. Most of them held this data closely; many even considered them secrets of state. Quetelet sought to obtain these numbers and then to publish them. He drew upon a broad European network of scientists to coax numbers from administrators and then published scores in his own journal.[19] In the age of the internet, journal publication may seem ridiculously slow—but it was a radical transformation in the public availability and circulation of data.

And he set to analyzing all this data. He adapted and simplified the data analysis of the astronomers, to find regularities in data on populations. A few regularities had been known for over a century—and were typically offered as evidence of Divine Providence organizing the world. The historian Kevin Donnelly has stressed how Quetelet sought to move from subjects like mortality, where human agency was limited, to "moral" domains like crime, where agency was paramount. As he gained access to data on crime and human physical characteristics, Quetelet came to recognize another kind of regularity, of the grouping of data around averages.

He was not content simply to note these regularities. He immediately granted them significance—and a form of reality. "This remarkable constancy with which the same crimes appear annually in the same order," Quetelet explained, "drawing down on their perpetrators the same punishments, in the same proportions, is a singular fact, which we owe to the statistics of the tribunals."[20] These statistical laws appeared to question human free will. They suggested that we do not each control our fate.

From Error Theory to Average Man

Quetelet argued that if we observe "moral phenomena" in large numbers of people, they come to resemble physical phenomena. The "greater the number of individuals observed, the more do individual peculiarities, whether physical or moral, become effaced, and leave in a prominent point of view the general facts, by virtue of which society exists and is preserved."[21] How to deal with lots of individual observations? Then as now, one could write lots of novels, and hope to yield something of the eternal human condition. Instead, Quetelet applied a new mathematical technology originally for dealing with abundant astronomical observations.

In his quest to build an observatory, Quetelet traveled to Paris where he learned how large numbers of astronomical observations, usually produced by many different people, could be converted into fairly certain knowledge about the positions of the stars and the planets in the night sky. If several people measure the position of a given star in the sky, the observed positions will vary from time to time and from person to person and from instrument to instrument.

The great mathematicians Pierre-Simon Laplace and Carl Gauss had shown that multiple astronomical observations of the same quantity tend to fall along what we often call the bell or normal curve. The center of the curve provides the location for a stellar body best supported by the evidence. Quetelet did something new with this astronomical technology for dealing with lots of data produced by different sets of eyes.[22] He applied this way of inferring to data about human beings—data like the incidence of crime or suicide rate or heights of a population. And then he made an enormously consequential jump—one not entirely justified by the science of the time.

If you and I made a bunch of observations of the position of a star over many nights, we'd be trying to ascertain a real value: the position of one star in the sky. Now, if we measured all the heights of the members of an army battalion, we could easily compute the average height. That average would be an abstraction from the data, not an attempt to measure something real, something out there. It's not like finding the position of the star.

Quetelet's flash of genius—whatever its lack of rigor— was to treat averages about human beings as if they were real quantities out there that we were discovering. He acted as if the average height of a population was a real thing, just like the position of a star—the number "objectively describes the population."[23] Despite "the fluctuation of numbers," he wrote, we know "that there's really a number whose value we seek to determine, whether it is the height of an individual . . . , or the right ascension of the polar star."[24] Just such numbers, Quetelet maintained, characterized the *homme moyen* or "average man" of a given population.

However ridiculous Quetelet's average man sounds in retrospect, creating measures that characterize entire populations is central to our policy: crime rates, GDP, IQ. If many of these are understood as abstractions, having no real existence, others, like innate intelligence of different ethnic or racial groups, are often interpreted as having some biological reality. And this treatment has radically affected access to education and resources, as well as entire accounts of the nature of human difference. We'll see in the chapters to come how other scientists created measures in the conceptual space that Quetelet opened up by characterizing a "race" objectively.

Quetelet's focus on discerning the "average man" provided a tool for characterizing what is most characteristic of a given society—or "race," in the nineteenth-century idiom. The philosopher Hacking explains, "A race would be characterized by its measurements of physical and moral qualities, summed up in the average man of that race."[25] Characterizing "races" in this way opened the space to understand comparative work of the differences between "races," between men and women and also to undertake developmental analysis, to understand differences of human beings over time and well as the development of individual human beings. All these facets were central to a new science of "man"—a new approach to understanding human nature scientifically.

The Science of "Man"

Treatises, leaflets, novels, bawdy poems—all claimed, across the European Enlightenment of the eighteenth century, to reveal the genuine nature of humanity. Were humans purely self-interested creatures, as Thomas Hobbes and many economists today might argue? Did they have pity? Were they fundamentally individualistic or social beings? They accrued scads of evidence, most of which appears anecdotal to us. As it did to Quetelet. "Experience alone can with certainty solve a problem which no a priori reasoning could determine," he wrote about discerning the nature of humanity. "It is of primary importance to keep our view of man as he exists as an insulated, separate, or in an individual state, and to regard him only as a fraction of the species. In thus setting aside his individual nature, we get quit of all which is accidental, and the individual peculiarities, which exercise scarcely any influence over the mass, become effaced of their own accord, allowing the observer to seize the general results."[26]

For Quetelet, knowledge of human nature doesn't come from armchair philosophers introspecting about the human condition or carefully described realistic novels capturing the nuances of an individual life. It will come from mathematical processes that will extract the "general results" characteristic of genuine human nature, not the accidents of this or that human being.

Governments tend to be good at recording major life events—birth and death and those moments where states and people interact. In the early nineteenth century, as now, those interactions often involve police, doctors, educators contending with what they take to be crime or deviance. Regularities had long been noted in birth and death data. Quetelet, for his part, emphasized the regularities to be found in data about crime.

> This remarkable constancy with which the same
> crimes appear annually in the same order, drawing
> down on their perpetrators the same punishments,
> in the same proportions, is a singular fact, which
> we owe to the statistics of the tribunals. In various
> writings, I have done my utmost to put this evidence
> clearly before the public; I have never failed annu-
> ally to repeat, that there is a budget which we pay
> with frightful regularity—it is that of prisons, dun-
> geons, and scaffolds.[27]

Even in the domain of moral actions, mathematical regularities appeared quickly. Following much of the best scientific thinking of his day, Quetelet held back from opining on the precise immediate causes of crime. In his typical way, he took the evidence of these regularities as evidence of the existence of something above and beyond individual human beings, a reality unifying groups of people. "Society," he controversially argued, "includes within itself the germs of all the crimes committed, and at the same time the necessary facilities for their development. It is the social state, in some measure, which prepares these crimes, and the criminal is merely the instrument to execute them." To understand the pace and increase of crimes, we need to understand the organization of society, not just of individuals. "Every social state supposes, then, a certain number and a certain order of crimes, these being merely the necessary consequences of its organization."[28] Here Quetelet applied a vision of the necessary causes from physics to the social world.

Indeed, he characterized his own achievement as showing that moral phenomena, when observed through data, resembled astronomical phenomena: "we thus arrive in inquiries of this kind, at the fundamental principle: that the greater the number of individuals observed, the more do individual particularities, whether physical or moral, become effaced, and leave in a prominent point of view the general facts, by virtue of which society exists and is preserved."[29] Understanding human society meant understanding these general facts, something made possible by the accumulation of ever-larger amounts of data about that society and its people.

Quetelet insisted that his discovery of the moral laws characterizing different societies should not dim hope, but rather point toward the possibility of improvement. That crimes emerge from the organization of society, he argued, should be "consolatory. . . . by showing the possibility of ameliorating the human race, by modifying their institutions, their habits, the amount of their information, and, generally, all which influences their mode of existence.[30]

Reification and Objectivity

Ian Hacking argues Quetelet dramatically transformed how we understand the world. With his work, "statistical laws that were merely descriptive of large-scale regularities" turned into "laws of society and nature that dealt in underlying truths and causes."[31] Finding the average isn't just a description of a group that we make up. For Quetelet, that average is something more real. The average captures something existing above and beyond each person, something about the group itself, something about how each individual group member acts.

Quetelet connected the study of social phenomena, which often fall into the normal curve, with the study of the variation of observations by human beings, which fall into the normal curve. The normal curve characterized underlying human variability, and, to the surprise of many, suggested an underlying lawlike behavior of humans in the aggregate. Every suicide may be a product of individual choice, but all the suicides in a year fall into knowable patterns. Putting state statistics into normal distributions helped make apparent properties characterizing the aggregate. Along with a number of other contemporaneous thinkers, Quetelet made the thinking of *society* as something more than merely a collection of individuals possible. In defense of individual agency and responsibility, Prime Minister Margaret Thatcher famously quipped in 1987, "There are individual men and women. . . . There is no such thing as society"[32]; yet Quetelet and his heirs repeatedly showed the numerical laws characterizing society. He called this *social physics*.

With his focus on the attributes of the average man, of the social, Quetelet inspired other sciences to focus on complex wholes that can be understood to have intelligible qualities even if we little understand all their component parts.

Historian Theodore Porter argues that Quetelet showed how "statistical laws can prevail for a mass even when the constituent individuals are too numerous or too inscrutable for their actions to be understood in any detail."[33] The source of much of the early mathematics for contending with observational data, physics, followed in the middle years of the 1800s the social sciences in embracing statistical law as appropriate for understanding the natural world. In turn, models and statistical procedures from physics came back to the social sciences.

Heading Off Revolution

Social physics in the manner of Quetelet didn't simply *describe*. It *prescribed*—told what to do, in a potent moral language of the improvement necessary required for modern societies engaged in industrialization and colonization. The "friends of humanity" must study the slow statistical transformations in society in order to pursue the gradual modifications desired.[34] Revolutions and disruptions need not apply. Theodore Porter remarks, social physics must "be recognized as a paean to social order in the spirit of gradualist liberalism."[35] If the central region of the bell curve was normal, its peripheries were evidently the realm of the pathological. Understanding populations was the prerequisite for attempting to take care of—and improve—deviant or pathological individuals caused in a lawlike way by their social settings. In this view, deviance took on a new role. Porter explains, "The implication of Quetelet's idealizations of the mean was that all deviations from it should be regarded as flawed, the product of error."[36] And this error could be known through science.

In his remarkable *The Taming of Chance*, Ian Hacking argues that "the average man led to both a new kind of information about population and a new conception of how to control them."[37] The heirs of Quetelet seized on his efforts to characterize people and races—and pushed them much further. Through statistics, Quetelet sought to improve human races. His efforts suggested the power of characterizing humans with new "objective" measurements and the need for improving social "average qualities of race." The heirs to his thought deepened and ran with these to craft a new scientific racism distinct from Quetelet's liberal improvement yet deeply indebted to it. The key figure was a cousin of Charles Darwin, one Sir Francis Galton, who turned Quetelet's work into a new "science of individual differences."[38]

Toward New Sciences of Human Improvement

In February 1891, that British advocate of Quetelet, Florence Nightingale, penned a long list of pressing policy questions in a letter to Galton. Reformers and their opponents had long debated policy without solid evidence. She asked for statistics to inform power, to answer such pressing issues as:

> The results of legal punishments—i.e., the deterrent or encouraging effects upon crime of being in gaol.
>
> . . .
>
> What effect has education on crime?

Speaking of India, she asked:

> Whether the peoples there are growing richer or poorer, better or worse fed and clothed? Whether their physical powers are deteriorating or not?[39]

The idea that we ought to use data and statistical analysis to answer such questions seems banal today. We have to understand how such forms of knowledge became obvious tools of governance, of power. It wasn't obvious to contemporaries. She called on Galton to "jot down other great branches upon which he would wish for statistics, and for some teaching how to use these statistics in order to legislate for and to administer our national life with more precision and experience."[40] Galton aimed, as we'll see, even higher, to administer nothing less than the intellectual and physical qualities of diverse races of the globe.

CHAPTER 3

The Statistics of the Deviant

I n 1915, a young graduate from Cambridge boarded a ship to return home to India. With him was a complete run of the most dynamic, the most exciting periodical of the day, focused on the science for understanding the present and building better futures through data. The journal, *Biometrika*, embodied a new data-driven approach to biological and social problems, including questions of race and heredity—a great furthering of the dreams of Quetelet. The graduate, Prasanta C. Mahalanobis, saw in them technologies both colonial and a future independent India could adapt, to know itself better in statistical ways, to guide its economic and social development. As the great institutionalizer of statistics in India, Mahalanobis transformed these methods while tempering at times their hubris. The new sciences Mahalanobis brought with him emerged from the intersection of data, Darwinism, and a crisis of confidence in imperial Britain. These sciences produced mathematical statistics, born amid British power—and great fears of decadence and decline.

Moral panics can create new sciences. Late nineteenth-century elite Britons were consumed with worries about their empire's degeneracy. "The time appears to have arrived," wrote Florence Nightingale in 1858, "when by the British race alone must the integrity of that Empire be upheld."[1] Declining birthrates among the elite, "limited population," alcoholism, and failures abroad all spoke to an empire in crisis. A new field, eugenics, and the statistics to support it provided one way not only to diagnose society, but to attempt to cure it—a way to make Britain great again.

A Crisis of the Modern Era

When the gentleman-scholar Francis Galton inspected his fellow men of Victorian Britain, he found them wanting: "We want abler commanders, statesmen, thinkers, inventors, and artists," he wrote in an article called "Hereditary Talent and Character." "The natural qualifications of our race are no greater than they used to be in semi-barbarous times," even though "the conditions amid which we are born are vastly more complex than of old." Modern civilization was all too much. "The foremost minds of the present day seem to stagger and halt under an intellectual load too heavy for their powers."[2] Genius was needed, but was in short supply. Education would never be enough, for there were simply not enough gifted men and women, not enough born geniuses, to confront the complexity of the times. England needed more geniuses, more people of extraordinary talent. They needed, Galton decided, to be bred.

The writings of Galton's illustrious if infamous cousin, Charles Darwin, offered a way forward.[3] In his *Origin of Species*, Darwin used the example of human breeding of domestic animals, such as show pigeons and pedigree dogs, to motivate his account of natural selection. Just as human breeders select features they desire, certain features of animals are selected for, as it were, over time in particular environmental niches. Humans, in Galton's estimation, underestimated their own power to affect their species. "The power of man over animal life," Galton explained, "is enormously great. It would seem as though the physical structure of future generations was almost as plastic as clay, under the control of the breeder's will." Not just physical traits, but equally the mind could be altered: "It is my desire to show more pointedly than—so far as I am aware—has been attempted before, that mental qualities are equally under control."[4]

Galton soon coined the term "eugenics" to describe the conscious effort to improve the quality of human beings—and national races of human beings in particular. Eugenics quickly became central to many left- and right-wing political programs across Europe, the United States, and the world. Racist to its core, to be sure, Galton's primary focus nevertheless was class. His suggestions often were whimsical, especially in comparison with the forced sterilizations and genocides associated with eugenic programs to come outside of Britain:

> Let us, then, give reins to our fancy, and imagine a Utopia—or a Laputa, if you will—in which a system of competitive examination for girls, as well as for youths, had been so developed as to embrace every important quality of mind and body, and where a considerable sum was yearly allotted to the endowment of such marriages as promised to yield children who would grow into eminent servants of the State.[5]

Unlike many philosophers and economists of his time, Galton was fundamentally anti-egalitarian.[6] "I object to pretensions of natural equality," he wrote. "I have no patience with the hypothesis occasionally expressed. . . . that babies are born pretty much alike, and that the sole agencies in creating differences between boy and boy, and man and man, are steady application and moral effort."[7] All people were not created equal, Galton insisted, and not all market agents had comparable mental capacities. Liberal political thought and liberal economics were just wrong in his estimation.

We associate eugenics and scientific racism with the far right, with Nazis. Things were otherwise around 1900. Many progressives as well as conservatives up to World War II saw science as capable of improving the human lot by improving the human race; indeed, one proponent noted, belief in eugenics offered "a perfect index of one's breadth of outlook and unselfish concern for the future of our race."[8] Statistical sciences were to replace the bigotries of old with evidence-based new sciences of human improvement: accounts of natural human hierarchies that moved easily from description to prescription.[9]

To improve the species, Galton needed to explore the wellsprings of talent and human excellence. Nurture was no explanation. Using biographical dictionaries of great men and women, Galton began investigating the density of talent and genius within families. In his long study *Hereditary Genius* of 1869, Galton studied prominent families and compared historical states with those around the earth. Despite the large number of his cases, his approach was intuitive and anecdotal. He generally argued that modern peoples were all lesser than ancient Greeks and that non-European peoples— he called them "races"—lesser than European ones.

While the approach in his book is largely anecdotal, Galton drew on Quetelet's normal curve to support his new ideas of ranking people and races. Quetelet used the normal curve to understand the qualities of a group as a whole. Galton used the same curve to understand variation within a group. Quetelet might seek the mean stature of Englishmen. Galton sought to understand the extremes of stature. His quarry was talent, not height, but he applied the same tools to one and other. The French sociologist Alain Desrosières explains, Galton used the normal curve as "a law of deviation allowing individuals to be classified, rather than as a law of errors."[10] What astronomers saw as errors to be eliminated, Galton saw as individuals to be ranked and classified. Every child getting test scores listing their percentile performance lives in the world Galton helped create.

And yet there was a major sticking point to all this investigation of excellence in distinguished families. Extremely tall people had tall children, but, on the average, those children were not as tall as their parents, reverting toward a population average height. Similar observations describe a wide range of human and animal traits. For breeders of animals—human or otherwise—this was a puzzle, one that would limit attempts to breed supposedly superior human beings. How to understand it? The answer would come from Galton's reworking of Quetelet's enthusiastic applications of the normal curve.

Why do offspring of tall parents tend not to be as tall as their parents, and more generally why do the attributes of a human group stay nearly constant over time? Galton came to explain both phenomena through what he called "regression," mathematically capturing the "tendency of that ideal mean filial type to depart from the parent type, 'reverting' towards what may be roughly and perhaps fairly be described as the average ancestral type."[11] With his statistical investigation, he discovered a powerful mathematical relationship between the amount of reversion of offspring and the extent of their parents' deviation from the mean. He not only showed the relationship to be linear, but also undertook what we would call today, thanks to Galton, the linear regression applied to

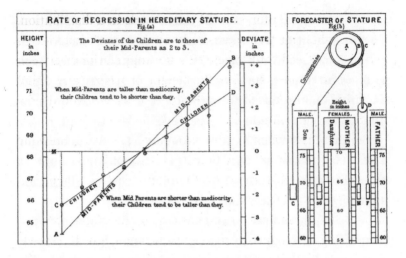

Francis Galton. "Regression Towards Mediocrity in Hereditary Stature." *The Journal of the Anthropological Institute of Great Britain and Ireland* 15 (1886): 246–63. Plate IX.

---

the data, finding the coefficients of a simple linear equation like $y = mx+b$.

Galton was modeling facets of the process of generation, so his initial work with reversion involved only treating the parental heights as the x's and only the children's heights as the y's, for he was looking at a unidirectional biological process. But he soon realized that his process of regression could be detached from its biological mooring and used on a vast array of data. In investigating the process of "reversion," Galton had unknowingly hit on a much broader concept, namely that of statistical regression.

Correlation and Data

Galton did far more than introduce a powerful new approach to modeling data and making predictions from data. Quetelet studied society. Galton studied individuals in a distribution. He wanted better techniques to know and rank individuals and to know and rank races. In studying relations between pairs of attributes, such as the height of parent and child, Galton also introduced "co-relation," or as we would now term it: the correlation.

While governments were producing an ever-increasing number of statistics, they failed to accumulate enough of the data that most interested Galton—detailed investigations of the "chief physical characteristics" of a wide selection of the population, qualities such as "Keenness of Sight; Colour-Sense; Judgment of Eye; Hearing; Highest Audible Note; Breathing Power; Strength of Pull and Squeeze; Swiftness of Blow; Span of Arms; Height, standing and sitting; and Weight."[12] So challenging was collecting this data that Galton set up an Anthropometric Laboratory at the International Health Exhibition of 1884 in South Kensington. The laboratory measured 9,337 people in seventeen ways. He explained that "periodical measurements" would be useful to families in tracking their individual development, and to "discover the efficiency of the nation as a whole and in its several parts." Such records "enable us to compare, schools, occupations, residences, races, &c."[13] The data produced would continue to be studied well into the twentieth century. Galton's anthropometry, historian of psychology Kurt Danziger explains, "defin[ed] individual performances as an expression of innate *biological* factors, thereby sealing them off from any possibility of social influence."[14]

Galton's style enabled a dramatic new approach to understanding human differences. Following Quetelet, analysis of data could reveal the commonalities and range of quantifiable human behavior and attributes. And following Galton, each individual could be placed and ranked within those ranges: the top 5 percent, the bottom 10 percent. Inspired by

Francis Galton, *Anthropometric Laboratory; Arranged by Francis Galton, FRS, for the Determination of Height, Weight, Span, Breathing Power, Strength of Pull and Squeeze, Quickness of Blow, Hearing, Seeing, Colour-Sense, and Other Personal Data* (London: William Clowes, 1884), 13.

Galton's work in observing large numbers of human beings, mental tests, for example, emerged from the effort to place each person amid the range of measured human capacities. And entire sciences of examining large numbers of "subjects" in statistical ways emerged in its wake. "A new method for justifying psychological knowledge claims had become feasible" with the work of Galton and his intellectual successor Karl Pearson, explains the historian Danziger. "To make interesting and useful statements about individuals it was not necessary to subject them to intensive experimental or clinical exploration. It was only necessary to compare their performance with that of others, to assign them a place in some aggregate of individual performances."[15] And it didn't take long for an approach to become big business. While pioneers like Galton struggled to get data at an adequate scale, a vast appetite for such inquiries would soon open, especially in the United States after the First World War.[16]

Above all, Galton revealed how surveying a mass of people makes recognizing—and targeting—the individual possible. Lots of data about lots of people allows scientists, marketers, militaries, spies to better know you—and target you. We live in such a world, where our individuality is quantified in reference to all other users of the internet, and where ad-serving algorithms exploit this quantification of difference to compete for our attention.

Institutionalizing Biometrics

The indefatigable Galton did not himself institutionalize his new statistical approach. He likewise did not have the mathematical skills to make it rigorous. Drawing on Galton's ideas and financial support, his intellectual heir, Karl Pearson, worked at both. A descendant of Quakers, freethinker, mathematician, socialist, feminist, and eugenicist, Pearson had a "grand vision, the creation of a statistical biology as the basis of effective eugenics and, concomitantly, the development of a mathematical statistics that could be applied to virtually all areas of human knowledge"—in the words of his biographer Theodore Porter.[17] Superior mathematical statistics would enable its expansion to the whole range of phenomena Quetelet dreamed of, an entire spectrum of social reform.[18] His field-building institutionalized eugenics and an imperious new statistical approach to social and political programs, with the help of patrons such as Galton and, of all things, the Worshipful Company of Drapers.

To do all this Pearson required data, labor to process that data, and new mathematics.[19] As he noted while giving a prestigious lecture, "the work is essentially the result of a co-operative investigation extending over a number of years, and depending upon a body of collaborators" who produced and analyzed "the extensive data on which my results entirely depend."[20] Pearson toiled for decades with a cadre of workers to bring his projects to fruition; a generation of great statisticians worked under and with him and changed how we all use data. Pearson ran multiple laboratories, including separate biometric and eugenic laboratories, with distinct projects, methods, staff, and funding.[21] With the help of two women assistants in particular, Alice Lee and Ethel Elderton, he amassed a wide range of data for a wide range of statistical investigations, and published results based on them, mostly in journals he founded and ran.

Getting data was hard work. In 1903, a plague pit was opened in London. Less than a week later, "one of my workers, Mr. S. M. Jacob, had with unwonted energy 'begged' the whole of the crania & skeletons" for Pearson's work.[22] Most data acquisition was more prosaic. Aiming to extend Galton's studies of the inheritance of physical and mental capacities, Pearson and his team placed requests in magazines read by headmasters and teachers asking them to record a multitude of observations on pairs of siblings and to rank them intellectually. They sent out 6,000 forms and got back some 4,000 from a wide range of schools (see illustration). "The absolute classification and tabling has been a work of great labour," Pearson explained, thanking a team of exceptional women: "Miss Alice Lee, D.Sc.; Miss Marie Lewenz, M.A., Miss E. Perrin, Miss Mary Beeton and Miss Margaret Notcutt" before noting that the "chief labour of computing has fallen upon Dr Alice Lee."[23]

Processing data was arduous and expensive, even with the help of new machines. Galton supported the Eugenics Lab, and in 1903, the Worshipful Company of Drapers granted Pearson £500 for his Biometric Laboratory, which allowed him to begin paying Alice Lee, who had previously

APPENDIX IB.

DATA PAPER FOR COLLATERAL HEREDITY INVESTIGATIONS.

B. SISTER-SISTER SERIES.   No. in whole series.
(Whole, not half sisters.)   (Not to be filled in.)

Please return this Paper to Professor KARL PEARSON, F.R.S., University College, London.

School:

Observer: No. in School Series

Date:

Place a cross against the class of each sister under as many headings as possible, except under III and VIII. Please read first the General Directions.

	ELDER SISTER.	YOUNGER SISTER.
Name		
Age		
District of Home		

I. PHYSIQUE:

	Very Strong.	Strong.	Normally Healthy.	Rather Delicate.	Very Delicate.	Athletic.	Non-Athletic.
ELDER SISTER ...							
YOUNGER SISTER ...							

II. ABILITY: (a) General Scale.

	Quick Intelligent.	Intelligent.	Slow Intelligent.	Slow.	Slow Dull.	Very Dull.	Inaccurate-Erratic.
ELDER SISTER ...							
YOUNGER SISTER ...							

(b) HANDWRITING:
(See Back.)

	Very Good.	Good.	Moderate.	Poor.	Bad.	Very Bad.
ELDER SISTER ...						
YOUNGER SISTER ...						

(c) WORK:

	Classics.	Modern Languages.	History.	Mathematics.	Descriptive Science.	Drawing.	Singing, Music.
ELDER SISTER ... Good at ...							
Best at ...							
Likes best							
YOUNGER SISTER Good at ...							
Best at ...							
Likes best							

(d) GAMES OR PASTIMES:

	ELDER SISTER.	YOUNGER SISTER.
Likes ...		
Good at ...		

III. HEAD MEASUREMENTS:

	Length.	Breadth.	Height.	a.	b.	c.	(a), (b), (c), Indices (not to be filled in).
ELDER SISTER ...							
YOUNGER SISTER ...							

IV. HAIR:

	Red.	Fair.	Brown.	Dark.	Jet Black.	Smooth.	Wavy.	Curly.
ELDER SISTER ...								
YOUNGER SISTER ...								

V. EYES:

	Light.	Medium.	Dark.
ELDER SISTER ...			
YOUNGER SISTER ...			

VI. RELATIVE CAPABILITIES: This is only to be filled in in those cases wherein the two sisters fall into the same class.

	Physique, stronger in	More Athletic.	Ability, greater in	Handwriting, better in	Hair, darker in	Eyes, darker in
ELDER SISTER ...						
YOUNGER SISTER ...						

VII. CHARACTER, ETC.:

	Noisy.	Quiet.	Self-conscious.	Unself-conscious.	Self-assertive.	Shy.	Conscientiousness. Keen.	Dull.	Popular.	Unpopular.	Quick.	Temper. Good-natured.	Sullen.
ELDER SISTER ...													
YOUNGER SISTER ...													

VIII. GENERAL REMARKS. Add here any striking features of resemblance or dissimilarity in the sisters.

ELDER SISTER ...	
YOUNGER SISTER ...	

[On the back of the Schedule spaces were arranged for samples of the handwriting.]

21—2

Data Paper for Heredity Investigations, in Karl Pearson, "On the Laws of Inheritance in Man: II. On the Inheritance of the Mental and Moral Characters in Man, and Its Comparison with the Inheritance of the Physical Characters." *Biometrika* 3, no. 2/3 (1904): 131–90, at p. 163.

undertaken extensive calculations for him on a volunteer basis, as well as collaborating with him. "Her duties included reducing data, computing correlation coefficients, creating bar charts . . . and calculating a new kind of statistic"—chi-squared—as well as supervising calculators male and female.[24] Calculation with machines became so central to the work at Pearson's laboratory that one visitor noted a "preoccupation with mastery of details of calculation" that could obscure the new mathematical statistics.[25] Much of this labor resulted in major collections of printed tables. It is hard to appreciate today just how essential such tables were as computational infrastructure for the growth of mathematical statistics.

While his women co-workers were often preoccupied with the tedium of calculating, Pearson also encouraged their higher-level work and often published with them. He argued, for example, "that Miss Elderton be no longer spoken of as a clerk, but be made a Francis Galton Scholar. She is quite capable of doing original work." Besides their contributions to statistics, they could become leaders in local social work. "It is most desirable that people trained in the Eugenics Laboratory should pass into work in public or municipal service of some type, as in dealing with mental defectives or invalid children, etc. We shall thus develop into a training school for practical eugenic work."[26] The most prominent of these women, F. N. (Florence Nightingale) David, named for the famous health reformer, went on to a storied career in statistics, including as professor in California.

Inheritance and Social Policy

What did all this labor with data prove? Intelligence was inherited, and Britain was losing the game of intelligence: "we are ceasing as a nation to breed intelligence as we did fifty to a hundred years ago. The mentally better stock in the nation is not reproducing itself at the same rate as it did of old; the less able, and the less energetic, are more fertile than the better stocks." This had major implications for social reform, as the problem wasn't schools but the breeding stock. "No scheme of wider or more thorough education will bring up in the scale of intelligence hereditary weakness to the level of hereditary strength." The only "remedy" is "to alter the relative fertility of the good and the bad stocks in the community."[27]

For Pearson, statistics was to be central for a new eugenic socialism necessary for a modernity both industrial and a conflict of races. If the goals were eugenic planning for a superior race, however, they were not trivial impositions of a racist and classist belief system onto data. Investigating skulls led Pearson and his collaborator Alice Lee to deny any reliable correlation between cranial size and intelligence, and to deny that skulls demonstrate the innate lower intelligence attributed to women.[28]

Eugenic statistics told tough truths: "we have failed to realize that the psychical characters, which are, in the modern struggle of nations, the backbone of a state, are not manufactured by home and school and college; they are bred in the bone; and for the last forty years the intellectual classes of the nation. . . . have ceased to give us in due proportion the men we want to carry on the ever-growing work of our empire, to battle in the fore-rank of the ever intensified struggle of nations."[29] The pressing political issues of the day required superior eugenical knowledge:

The whole problem of immigration is fundamental for the rational teaching of national eugenics. What purpose would there be in endeavouring to legislate for a superior breed of men, if at any moment it could be swamped by the influx of immigrants of an inferior race, hastening to profit by the higher civil-isation of an improved humanity? To the eugenist permission for indiscriminate immigration is and must be destructive of all true progress.[30]

Like Galton, Pearson argued that the "struggle of nations" was simply too important to rest on false eugenical science: that struggle required better science.

A Superior Science of Big Data

The social and biological sciences needed a remaking based in mathematics and in the production of data: "the loose qualitative or descriptive reasoning of the older biologists must give way to an accurate mathematico-statistical logic. The trained biologist may discover and tabulate facts, much as the physicist does today, but it will need the trained mathematician to reason upon them. The great biologist of the future will be like the great physicist of to-day, a mathematician trained and bred."[31] Many contemporary biologists disagreed, needless to say. Pearson extolled large-scale data collection and analysis rather than small-scale laboratory and experimental work.

What was true in biology was even more true in politics. Pearson noted with irritation the ease with which people opine on social questions: "every politician, every platform orator, who would hesitate to express even his opinion regarding a question in astronomical physics or cytology is ready with a decisive answer to each social problem that arises." But social problems were far harder than astronomical ones. "Social problems needed scientific answers. Every social problem belongs to a class embracing the hardest of all problems—it is vital not physical, it is biological, it is medical, it is statistical. It needs not less but far more investigation for its solution than any academic physical or biological problem."[32] Pearson's laboratories offered models for organizing political and social order along these new scientific lines.[33]

Correlation, Not Causation

Correlation, we are ever taught, doesn't equal causation. And to Galton's intellectual heir, that's why it was so exciting. Karl Pearson explained that he realized there was a "category broader than causation, namely correlation, of which causation was only the limit." Now more sciences could be made mathematical: "this new conception of correlation brought psychology, anthropology, medicine and sociology in large parts [sic] into the fields of mathematical treatment."[34] Correlations were particularly attractive in looking at sets of data with no clear causal relation. In studying evolution, correlation helped understand the processes of evolution without providing knowledge of its causes. Pearson believed fertility was strongly correlated with lower intelligence, lower morals. Correlation was essential, for example, to understand the reproductive policy a nation should follow if it was not to decline. Correlation, he proclaimed late in life, "has not only enormously widened the field to which quantitative and therefore mathematical methods can be applied, but it has at the same time modified our philosophy of science and even of life itself."[35]

Most of statistics in the twentieth century focused centrally upon causation, as we will see in the chapters to follow. But much of our current data revolution involves the reemergence of correlation as the most important tool in commerce, spycraft, and science. Whether in finding correlations or claiming expertise about the social world, a Pearsonian spirit pervades the data sciences.

New Data-Driven Racisms

From our point of view, all these figures appear largely backward racists and classists. And they were. Yet they were not hidebound traditionalists or conservatives. To the contrary: their science was central to part of their progressivism, to how they proposed to study social difference, and to foster a national unity that they believed to be undergirded by the best knowledge of their day. These new sciences would disrupt the conceptual foundations for the social order—even if they didn't ultimately change this order very much. Showing how radical technical *disruptions* often serve to reinforce existing inequalities will be a theme throughout this book.

Eugenicists saw few of their favored policies adopted as quickly as they would have liked, so some historians have dismissed the significance of the movement. Historian Robert Nye explains, "the long-term importance of a eugenics discourse in England was the way it transformed a narrow class outlook into a matrix of biomedical concepts claiming to represent the interests of the whole society, and which became an irresistible perspective for generations of educated Britons."[36] Eugenical ideas became default ideas for many in the educated classes; in Britain concerns with class predominated; in the United States race figured prominently. Eugenical ideas also shaped policy in Nazi Germany, with genocidal results.

Biometry, Race, and the Problems of Modern Society

"If modern civilisation is distinguished from all other civilisations by its scientific basis," Brajendranath Seal explained, "the problems that this civilisation presents must be solved by the methods of Science." In this opening address to the

1911 First Universal Races Congress, with W. E. B. Du Bois in attendance, Seal argued that the solutions to the pressing problems of race in the modern world required new sciences of humanity—not the old humanistic or philosophical methods of an Aristotle or a Machiavelli, but the new biometric sciences. "A scientific study of the constituent elements and the composition of races and peoples, of their origin and development, and of the forces that govern these, will alone point the way to a settlement of inter-racial claims and conflicts on a sound progressive basis," in the divided US, the restive British Empire, and the rest of the world.[37]

Embracing a eugenics program, Seal noted, the "study of genetic conditions and causes, of the biological, psychological, sociological forces at work, which have shaped and governed the rise, growth, and decadence of Races of Man, can alone enable us to guide and control the future evolution of Humanity by conscious selection in intelligent adaptation to the system and procedure of Nature."[38] And yet Seal distrusted the usual division of humans into races, and called for biometry to delineate properly the divisions of humanity, based on data. Seal was imbued with the approaches of Galton and Pearson: we must "adopt biometric methods in studying characters and variations," distrust averages, as "the range of variations in a character is as important an index as the character itself."[39]

A few years later Seal told Prasanta Mahalanobis, "You have to do work in India similar to that of Karl Pearson in England." In building institutions and pursuing biometric investigations, Mahalanobis did so. He brought the biometrical program to India, developed and challenged Pearson's methods, and founded mathematical statistics in India.[40]

Committed both to the acquisition of biometrical data and the development of ever more rigorous investigation of it, Mahalanobis came ultimately to turn highly problematic colonial data produced by the English into potent forms of nationalist knowledge as India secured its independence. In time, he made colonial data serve the new postcolonial Indian state.[41]

Following the aspirations of Seal, Mahalanobis sought techniques for discerning racial and caste mixture of various populations. Today he is best known for a measure of distance used in statistics that he first developed, "caste difference," as an alternative to Pearson's approach to the scientific study of racial difference. Unlike many of the racial theorists of their time, Seal and Mahalanobis stressed slow but real transformations over time. In his 1925 study of the Anglo-Bengals, Mahalanobis discerned dramatic but intelligible change. Caste had some transitory reality he claimed but "caste-synthesis" was well under way. "Intermixture within the province has gone on slowly and steadily even if imperceptibly and a larger Hindu Samaj has evolved which is not only not identical with the traditional society of Vedic or classic times but is in many respects even antagonistic."[42] The data analysis revealed the slow biological creation of a new Indian nation with real biological unity out of caste and sectarian division.

Aiding his approach to quantifying caste were powerful new tools for examining correlations among social groups at great scale. His new empirical techniques, he argued, both revealed this slow unification but equally the diversity of castes and tribes in India. In a massive data analysis, Mahalanobis and his collaborators undertook a data-driven clustering of castes and tribes in Uttar Pradesh.

Producing these analyses involved not just a team of human calculators, but also the use of "Mallock's Machine" housed in Cambridge in the United Kingdom.[43] Both in their empirical approach focused on calculating correlations at huge schools and in the use of new calculating devices, Mahalanobis and his team were doing data science long before data science. They found clear distinctions between

Brahmins, artisans, and tribal groups. And yet, for all the power of his techniques, Mahalanobis recognized the limits of the implications of these numerical differences. "To make further progress, it is necessary to take into consideration the social and cultural history of the tribes and castes, that is, the known ethnological evidence."[44] The failure to turn to such expert knowledge would plague—and indeed plagues to this day—too much data-driven science. No matter how powerful the algorithm or extensive the data, if one fails to embed this data analysis within broader forms of knowledge, scientific and humanistic alike, that so-called knowledge should be seen as incomplete at least, dangerous at worst.

Yearning for Causes—of Racial and Class Difference, for Example

Reviewing the 1911 First Universal Races Congress where the Indian intellectual Brajendranath Seal envisioned the coming together of nations, the American sociologist and delegate W. E. B. Du Bois drew out the most significant take-aways. "History illustrates these truths," he wrote on his notes, before quoting a distinguished speaker. "If we find an immense difference between the mind of some race" in Africa "and that of European race, we must seek the cause not in any difference of national qualities," but in external conditions. "It is not a difference of mentality in the race, but a difference of instruction, the same difference that we find to a greater or less extent, between the various classes of one and the same race or the different periods of its history."[45] Race and class differences must not be taken for granted, and differences in current intelligence ought not to be ascribed based on existing differences.

CHAPTER 4

Data, Intelligence, and Policy

D ecades before the Nazis built a state around their racial science, an American insurance employee claimed to have data demonstrating the inherent superiority of the "Aryan race." In 1896, a German immigrant to the United States, Frederick Hoffman, published a brutal portrayal of Black Americans in the second half of the nineteenth century, under the auspices of the American Economic Association. To Hoffman's eyes, the data annihilated the smug egalitarianism of liberal figures such as John Stuart Mill. Mill stressed the equality of the sexes as well as races. Data, according to Hoffman, proved indubitably otherwise. And government and corporate policies must take heed of the science of inequality, he insisted, whether in European colonies or in the American South.

"Only by means of a thorough analysis of all the data that make up the history of the colored race in this country," wrote Hoffman, "can the true nature of the so-called 'negro problem' be understood and the results of past experience be applied safely to the solution of the difficulties that now confront this country."[1] Three hundred pages later, the author turned to what the data showed, "It is not in the conditions of life, but in race and heredity that we find the explanation of the fact to be observed in all parts of the globe, in all times and among all peoples, namely, the superiority of one race over another, and of the Aryan race overall."[2] Hoffman didn't limit himself to Blacks in the United States. The data, he maintained time and again, showed conclusively that colonized people worldwide, like Blacks in the US, had higher mortality and lower standards of living, not due to any environments or societal conditions, but, said Hoffman, to their innate inferiority.

In the second half of the nineteenth century, the old racisms of the past sought new legitimation in sciences grounded in the new fields of anthropology, sociology—and statistics. And these racial sciences provided cover for the creation of the broad array of laws and practices disenfranchising Blacks in the US known as Jim Crow. So-called "race realists" of today continue this heritage of dressing up prejudice and systemic inequality in scientific garb.

Hoffman was a hired gun.[3] The Prudential Insurance Company had employed him to fend off anti-discrimination laws forbidding insurers to charge Black clients more—honoring the promise of equal protection of the Fourteenth Amendment to the US Constitution. His much-celebrated work purported to show to his employers that Blacks were simply uninsurable. The data demonstrated, he claimed, that they were failing in the Darwinian struggle for existence. Drawing upon a wide array of data sources, Hoffman turned the task of illustrating the different mortality rates of Blacks and whites into a supposedly scientific statement of racial hierarchy, with warnings against the dangers of racial mixing added for good measure.

Critics, notably scholars of color, demolished Hoffman's reasoning. In a devastating review of this work, the sociologist—and later cofounder of the NAACP—W. E. B. Du Bois tore apart Hoffman's choice of data, stressed the limitations of the data for drawing general conclusions, and, above all, showed how many claims made about race applied to working classes and recent immigrants of all races. Far from proving the essential differences between Blacks and whites, the data provided an index of the socioeconomic differences between them. Du Bois noted that the author "has by no means avoided many fallacies of the statistical method. This method is after all nothing but the application of logic to counting, and no amount of counting will justify a departure from the severe rules of correct reasoning." As to "race traits" and the "conditions of life," Du Bois noted, "it would seem incumbent on him . . . to prove these race traits after being held in abeyance for at least a century, first took decisive action in the decade 1880 to 1890."[4]

And yet, even as he denied the inferiority of Blacks as a whole, Du Bois embraced eugenicist views that all races had their share of natural "degenerates."[5]

DU BOIS'S STATISTICAL analysis was correct, but his analysis was largely disregarded; Hoffman was in truth ridiculous, but powerful interests had strong incentives to believe otherwise. Methods of auditing and analyzing algorithmic decision making are powerfully illuminating, whether in 1900 or 2022, but are too often without effects without some configurations of power or of publicity to make them powerful. It isn't enough to be right, as Du Bois was.

Hoffman's statistical analysis became, as in the words of historian George Frederickson, "the most influential discussion of the race question to appear in the late nineteenth century." As his employer wanted, Hoffman's work justified refusing to insure African Americans early in the twentieth century; the work, and others like it, gave a scientific veneer to the creation of the entire apparatus of discrimination and disenfranchisement.[6]

And far from dismantling socioeconomic inequality, Hoffman's documentation helped to reinforce it. Categorically denying life insurance to any entire class of people had systemic effects in deepening impoverishment from generation to generation. Rather than dismantling its causes, the inequality was essentialized, treated as natural. New statistical approaches radically altered how Blackness was conceived. The historian Khalil Gibran Muhammad explains, in the early twentieth century, "Blackness was refashioned through crime statistics." Through "racial criminalization," he argues, Blackness became "a more stable racial category in opposition to whiteness"—especially as previously marginalized immigrant groups—the Italians and the Poles—lost their fearsome reputation.[7]

Statistics doesn't simply represent the world. It transforms how we categorize and view the world. It transforms how we categorize others and ourselves. It changes the world. And, as we'll see, contemporary data science does this—at hyperspeed.

Hoffman tried to make biology explain inequality—to naturalize its quality—to make it a thing. He was no great statistician. Right around the turn of the twentieth century, better statisticians provided new ways of understanding— and justifying—human difference. They, too, were prone to the errors Du Bois had espied in Hoffman.

Yule: On the Causes (er . . . Correlations) of Poverty

Despite vast influence, Hoffman was statistically underpowered. In making arguments for policy, he did not draw upon the new powerful tools of Galton and Pearson to predict and to prescribe. A fine mathematician and sometime employee and colleague of Karl Pearson named Udny Yule took those new tools beyond the study of human difference and applied them to vital social issues of the day. In keeping with societal anxieties of his era, Yule turned the latest technological development, regression with multiple variables, to the causes of the ebb and flow of poverty.

What drives poverty? A lively debate in Britain toward the end of the nineteenth century concerned what policies increased or discouraged poverty. Did direct support encourage more poverty? In 1834, Parliament had enacted Poor Laws, to discourage the population from "wanting" to be poor, by forcing anyone deemed capable to work in workhouses with deliberately harsh conditions. "Out-relief," the direct granting of funds for survival, was to be forbidden to all able-bodied adults and their families, in favor of "in-relief" for those working in such workhouses. But how did these forms of relief affect poverty? Did the tough-love approach curtail poverty, as proponents then as now often argue? Was there data and science to back up what had long been a largely moral argument?

In the late nineteenth century, statisticians in England sought to use data to answer these questions. We should not lose sight of how *odd* this remained at the time. Today we expect policymakers to be data-driven—or at least to pretend to be. Although some critics today contest the data undergirding the scientific consensus about vaccines and global climate change, generally, we expect to turn over substantial technical facets of our democratic decision-making processes to experts armed with data and means for analyzing it. We collectively give them this power.

Deferring to scientific expertise in questions of policy was not an obvious move. Such deference turns a policy question into a scientific one, one thought to be outside partisan rancor or philosophical debate. In the words of the historian Alain Desrosières, "a political problem" was translated "into an instrument of measurement that allowed arbitration of a controversy."[8] As we saw above, Hoffman sought to provide a scientific grounding for racial discrimination—and elimination.

In the 1890s, both sides in the English poverty debate began drawing on statistics. Reformers such as Charles Booth saw themselves as advocating a scientific approach to major questions of policy, an approach untainted by traditional political divisions and moral views. "Science must lay down afresh the laws of life," he wrote. A scientific, not religious, approach will "lead us on till we find the true solution of the problem of government."[9] Booth first articulated a vision of a poverty line, distinguishing those who could minimally care for a family and those unable to do so. In 1894, Booth released *The Aged Poor in England and Wales*, an epochal work of social description, chock-full of data and tables. Booth's approach depended on shoe leather. His vision aimed to survey all of London, but to do so through finely detailed "local knowledge" collected with the help of a team.

Based on the data, he ended the book with a series of politically significant claims. In particular, he denied the tough-love belief that being too generous with "out-relief" went along with higher poverty: "The proportion of relief given out of doors bears no general relation to the total percentage of pauperism."[10] Booth's statistical procedures soon came under a sustained attack by Udny Yule.

Yule applied the techniques Galton and Pearson had developed for biology to questions of policy. He transformed regression from a tool for studying inheritance to one for fitting lines to data in the investigation of causality.[11] Every psychologist, every political scientist, every economist running a regression is working in the vein of Yule, who applied the new technologies of data analysis to make policy-relevant claims about the social world. The new experts needed more than data: they needed powerful analytical technologies, to represent, to predict—and to prescribe.

In 1899 Yule published "An Investigation into the Causes of Changes in Pauperism in England," where he explored the relation between public assistance and poverty. Yule's answer? The opposite of Booth's. Yule claimed the data showed that financial assistance causes poverty to increase. Yule sought to reveal the *causes* of changes in poverty. This approach allowed one to move from interpreting a regression as a *prediction* to as a *prescription*; interpreted as knowledge of cause, we could set out policy prescriptions.

But how do you figure out what causes something? Causation was old-school, dead in the eyes of Yule's mentor Karl Pearson. Pearson believed knowledge of causes was *impossible*. Pearson instead celebrated the power of correlation to replace our yearning for causal knowledge:

It is this conception of correlation between two occurrences embracing all relationships from absolute independence to complete dependence, which is the wider category by which we have to replace the old idea of causation. Everything in the universe occurs but once, there is no sameness of repetition. Individual phenomena can only be classified, and our problem turns on how far a group or class of like, but not absolutely same, things which we term "causes" will be accompanied or followed by another group or class of like, but not absolutely same things which we term "effects."[12]

Initially following Pearson, Yule ultimately sought to overcome this way of limiting human knowledge. The temptation for causality led Yule to push harder: to create new math and new technologies for thinking through policy-relevant data.

Yule recognized the great philosophical dangers involved: "The investigation of causal relations between economic phenomena . . . offers many opportunities for fallacious conclusions." The complexity of the social and economic realms did not allow for the massive simplifications of physics. A statistician could not, he explained, "make experiments for himself," so "he has to accept the data of daily experience, and discuss as best he can the relations of a whole group of changes." Unlike a physicist, he cannot "narrow down the issue to the effect of one variation at a time. The problems of statistics are in this sense far more complex than the problems of physics."[13]

Statistics needed new tools for investigating the complexity of society and revealing the causes of social ills such as growing poverty. How do we measure it in different places and times?

To begin answering these questions, Yule drew on Galton's and Pearson's tools. They had focused their tools largely on biological data: the relationships among generations of animals and among the body parts of any given animal, including, notoriously, human beings. Drawing particularly on the tools of regression, Yule turned them on to economic phenomena. Yule eventually argued that observational data could be used to infer causality and structure policy choices, when combined with background knowledge.

Good science would need to face up to the complexity of economic change—in this case, the "various causes that one may conceive to effect changes in the rate of pauperism."[14] According to Yule, possible causes included:

1. Changes in the method, or strictness, of administration of the law.

2. Changes in economic conditions, e.g., fluctuations in trade, wages, prices, and employment.
3. Changes of a general social character, e.g., in density of population, overcrowding, or in the character of industry in a given district.
4. Changes more of a moral character, illustrated, for example, by the statistics of crime, illegitimacy, education, or possibly death rates from certain causes.
5. Changes in the age distribution of the population.[15]

The first category is of particular interest, Yule said, because, then, change "may be comparatively rapidly effected by the direct action of the responsible authorities."

But how to investigate?

From Correlation to Cause

Drawing upon the tools of Pearson and Galton, Yule discerned that out-relief and pauperism were, in fact, strongly *correlated*: "the rate of total pauperism is *positively* correlated with the proportion of out-relief given, i.e., high average values of the former correspond to high average values of the latter. The method used seems to leave no room for doubt."[16]

Booth had argued the opposite, drawing on the analysis of several examples. Yule criticizes Booth's confusion of examples for the whole: "it is extremely regrettable that a statist of Mr. Booth's standing should have given so many examples of the fundamental mistake of founding general conclusions on particular instances."[17]

And yet Yule initially insisted on taking care in understanding this argument: his claim "does not say either that the low mean proportion of out-relief is the cause of the lesser mean pauperism or vice versa." He explained, "To be quite clear, I do not mean simply that out-relief determines pauperism in one union, and pauperism out-relief in another, so that you cannot say which is which on the average: but I mean that out-relief and pauperism mutually react in one and the same union."[18]

Yule found correlation by itself inadequate. How could it guide policy? But how to overcome the dangers inherent in reasoning poorly from correlations?

A standard regression would take the form:

Change in pauperism =
A + B × (change in proportion of out-relief)
Where A and B are constants.

What's the problem with this? "The association of the changes of pauperism with changes in proportion of out-relief might be ascribed either to a direct action of the latter on the former, or to a common association of both with economic and social changes."[19] In other words, they might correlate just because a common cause makes them move together.

Yule sought to tame this dragon by building into his regression a selection of other features:

Change in pauperism=
a + b ×   (change in proportion of outrelief)
+ c ×   (change in age distribution)
+ d ×
+ e ×   } (changes in other economic, social,
+ f ×   ) and moral factors)

With this Yule sought to isolate the causes—to show, for example, that changes in age distribution are not the common cause.[20] The historian Stephen Stigler writes that he "used the regression equation as a device both to uncover the relationship he sought and to allow for potentially influential changes in the other variables he had at hand."[21] He proceeded methodically, until he came to believe he had exhausted the possible alternative hidden causes. "Unless, and until, then, it can be shown that some other quantity whose changes are closely correlated with changes in out-relief ratio can account for this observed association, there is no alternative to considering the result as indicating a direct influence of change of policy on change of pauperism."[22]

Undertaking this task required substantial calculation, and Yule availed himself of a Brunsviga mechanical calculator. "Without such mechanical aids to calculation," he noted, "I could scarcely have undertaken the present work."[23]

Yule concluded his paper: "Changes in rates of total pauperism always exhibit marked correlation with changes in out-relief ratio, but very little correlation with changes in population or in proportion of old in the different unions." Connecting with policy, he noted: "It seems impossible to attribute the greater part, at all events, of the observed correlation between changes in pauperism and changes in out-relief ratio to anything but a direct influence of change of policy on change of pauperism."[24] And yet . . . Yule never really surmounted the concerns around the dangers of conflating correlation and causation. And neither have many of the disciplines in his wake. "The investigation of causal relations between economic phenomena," he noted, "offers many opportunities for fallacious conclusions."[25] True that, as Yule's sharpest early critic, the economist Arthur Pigou, noted:

It observed that, in the various unions of this country, various lines of policy are being and have been pursued, and proposed by statistical reasoning to demonstrate *a posteriori* that one line of policy has better economic effect than another.[26]

"The fundamental objection to" this form of statistical reasoning, Pigou argued, "is that some of the most important influences at work are of a kind which cannot be measured quantitatively, and cannot, therefore, be brought within the jurisdiction of statistical machinery, however elaborate."[27]

It's not hard to postulate an underlying cause that would account for the correlation between out-relief and pauperism. And real knowledge on the ground easily supplied one. Based on that knowledge, Yule's 1909 opponent Pigou suggested that "better administration" underlay increases of pauperism and of out-relief. "In view of this circumstance, what more natural than to suggest that the observed correlation between the out-relief ratio and the pauperism percentage is due, not to any direct causal connection, but to the fact that both have been caused by the character of the general administration?"[28]

This critic was no Luddite, however, but a careful thinker about the dangers of reasoning from correlations. The difference can be seen in the diagram. Yule claims the first is true, while his opponent posits an alternate cause.

A critic writing nearly a century later, the statistician David Freedman, celebrated the knowledge gained by "shoe leather" in the face of statistical modeling: "statistical technique can seldom be an adequate substitute for good design, relevant data, and testing predictions against reality in a variety of settings."[29]

For all its technical prowess, Yule's approach had no answer as to how to use mathematics alone to disambiguate correlation and causation. Unlike many of his later followers, Yule understood this. Despite adding the word "Causes" to the title of his paper, a buried footnote provides his epistemic escape: "strictly speaking, for 'due to' read 'associated with.' "[30]

While Yule's work had little immediate practical effect in debates about Poor Laws, his techniques would become

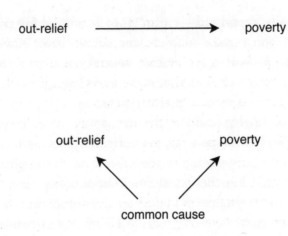

Alternate potential causes for increase in out-relief and pauperism. Our diagram, inspired by David Freedman, "From Association to Causation: Some Remarks on the History of Statistics," *Statistical Science* 14, no. 3 (August 1, 1999), 248–89.

central as one discipline after another sought scientific status: first economics, then psychology, then political science would all come to have multiple regression as a foundational technique essential to their claims to expertise. Although Yule's analysis didn't affect policy in his time, analysis of his kind has structured our lived reality for generations. Despite the logical and evidential problems, regressions remain a dominant tool in the social sciences and sciences of policy, and indeed often serve as a necessary sign of something being a scientific analysis.

Design and Proxies

This nineteenth-century debate is fundamentally about *poverty*. Like prosperity, poverty cannot be measured directly. So, anyone wishing to quantify it needs to choose something more easily measured—a proxy—to stand in for poverty. Scientists must make such choices all the time. While necessary, such choices are neither neutral nor unproblematic. They are design choices that make knowledge possible—but also subject to dramatic misinterpretation.

In the English debate, the key proxy for poverty was *pauperism*. Unlike poverty, pauperism is an *administrative* category. It's not a quality of people so much as the way a government classifies them. Administrative categories provide ways of classifying people using set definitions that bureaucracies can administer at great scale. And they produce data sets open to analysis. Administrative categories are powerful conventions, necessary to analyze society, but are not truths of nature, just existing to be found. The French historian Desrosières explains that an object like pauperism "exists by virtue of its social codification, through the reification of the results of an administrative process with fluctuating modalities." Reification is the fancy word for thinking that ideas are things—literally it means making a thing out of an abstraction about real things. It's a dangerous mistake, yet a constant danger in using statistical work in thinking through social, political, and commercial problems. "It is this slippage from the process to the thing," Desrosières writes, "that made Yule's conclusion so ticklish to interpret."[31]

Reification involves the fundamental vice of claiming existence for that which is a convention useful to us. And perhaps nowhere has the dangers of reification had such baleful effects as in the study of intelligence and its relationship to race.

Birth of Intelligence Testing

Among the many duties of the statistician Prasanta Mahalanobis was analysis of the management of schools and colleges. Like administrators across the world then and now, he turned to intelligence testing as part of the process of admission to educational institutions. Like other intelligence testers, he correlated academic success with a test originally based on academic success (IQ) and discovered they were well correlated. Creator of the first Bengali-language intelligence tests, Mahalanobis, however, was careful not to do what many of his contemporaries did with abandon. Despite his decades of work investigating castes and tribes in India, he offered no grand proclamations of IQ differences among the castes, and in particular claims that variations in innate intelligence explained—and justified—differences in social and political status. He didn't make a test into a claim about the nature of intelligence. He didn't use it to justify a natural hierarchy. The scholar Shivrang Setler has shown how he treated the tests as practical instruments.[32]

Many of his contemporaries were far less restrained. They took correlations between tests and measures of success and imputed cause. They reified proxies into something for which they had no plausible mechanisms. And worse, and at great scale, they called for and offered scientific justification for eugenics programs ranging from anti-immigration measures to forced sterilization.

In 1904, an English psychologist revealed a surprising result: the study of Latin and Greek offered good credentials for leaders, even in a modern age of steam engines, telegraphs, and railroads. Competency in Latin and Greek correlated most with what the author dubbed "General Intelligence."

Instead of continuing ineffectively to protest that high marks in Greek syntax are no test as to the capacity of men to command troops or to administer provinces, we shall at last actually determine the precise accuracy of the various means of measuring General Intelligence, and then we shall in an *equally*

Activity.	Correlation with Gen. Intell	Ratio of the common factor to the specific factor	
Classics,	0.99	99 to	1
Common Sense,	0.98	96	4
Pitch Dis[crimination].,	0.94	89	11
French,	0.92	84	16
Cleverness,	0.9	81	19
English,	0.9	81	19
Mathematics,	0.86	74	19
Pitch Dis. among the uncultured	0.72	52	48
Music,	0.7	49	51
Light Dis.,	0.57	32	68
Weight Dis.,	0.44	19	81

positive objective manner ascertain the exact relative importance of this General Intelligence as compared with the other characteristics.[33]

In a remarkable piece of experimental psychology, the author, one Charles Spearman, illustrated that a range of different cognitive and sensory abilities were strongly correlated with each other. And then he reasoned that they could all be understood in terms of an underlying form of general intelligence, called g.

In Spearman's account, someone who excelled in classics was intelligent, not due to the study of classics, but because excellence in the study of classics was the *most* correlated with other forms of excellence. Excellence in classics best indicated a high level of innate intelligence, or g.

In creating g, Spearman reduced a large number of measured aptitudes to a single value that could be ranked. This work rested on novel mathematics: he showed how to take many variables and discern the underlying or "latent" factors that underlie their changes and correlation, a process known as factor analysis, and central to modern statistics. This technique was potent if one was interested in ranking human beings, since it took a rich portrayal of many facets of each person and reduced them to something measured by a single underlying factor. Such techniques are central in our algorithmic world.

Spearman, and many who followed him, went one crucial step further. They turned this latent factor—an abstraction from a correlation among abilities—into a thing. They "reified" it: this correlation was transmogrified into a real thing humans possess, a general intelligence. And this intelligence was, by and large, he believed, inherited.

In his eugenics program, Galton had assumed that the best people had higher natural ability, but he had no way of measuring that ability directly.* Instead, he asserted that

---

* "By natural ability, I mean those qualities of intellect and disposition, which urge and qualify a man to perform acts that lead to reputation. I do not mean capacity without zeal, nor zeal without capacity, nor even a combination of both of them, without an adequate power of doing a great deal of very laborious work . . . nature which, when left to itself, will, urged by an inherent stimulus, climb the path that leads to eminence, and has strength to reach the summit—one which, if

reputation provided a good proxy for gauging natural ability. A superior procedure was clearly needed. Spearman provided the missing techniques for measuring the intelligence of anyone and, crucially, for placing them into a hierarchical order of the most intelligent (or best) to the least. As one historian explains, "The science of individual differences was invented by Francis Galton, systematised by Karl Pearson, and applied to psychology by Charles Spearman."[34]

And with these techniques for identifying intelligence, Spearman had finally allowed psychology to make the jump to a true science. In 1923, he explained, "we must venture to hope that the so long missing genuinely scientific foundation for psychology has at last been supplied, so that it can henceforward take its due place along with the other solidly founded sciences, even physics itself."[35]

Discussing the US Immigration Act of 1924 (the Johnson-Reed Act), a notorious restriction of immigration, Spearman explained,

The general conclusion emphasized by nearly every investigator is that, as regards "intelligence," the Germanic stock has on the average a marked advantage over the South European. And this result would seem to have had vitally important practical consequences in shaping the recent very stringent American laws as to admission of immigrants.[36]

Despite this Spearman was cautious, noting the differences in social conditions around education:

---

hindered or thwarted, will fret and strive until the hindrance is overcome, and it is again free to follow its labour-loving instinct." Francis Galton, *Hereditary Genius: An Inquiry into Its Laws and Consequences* (London: Macmillan, 1869), 37, 38.

There has been found a large body of evidence that races do differ from one another, at any rate in respect of *g*. . . . Nevertheless such racial differences, even if truly existing, are indubitably *very* small as compared with those that exist between individuals belonging to one and the same race.[37]

Those inspired by Spearman proved far less restrained in their conclusions. So less restrained that they didn't let a little academic fraud get in the way of some racist science, as numerous other authors, such as Stephen Jay Gould, have long demonstrated. Galton's followers, many far more eugenically rabid than Spearman, eagerly adopted his techniques.[38]

Karl Pearson worried that his fellow eugenicists might let their desire for proofs of the inheritance of traits get in the way of more careful inquiry. He was right. Pearson attacked Spearman, a colleague at University College London, for failing to prove his hypothesis of general intelligence:

The nature of the non-overlapping mental abilities should be selected by psychological consensus before the tests are made, . . . trained computers should be employed, and if possible a more adequate mathematical theory of the whole subject developed. Then we might have a better chance either of dismissing the whole theory, or showing that it was worthwhile spending further energy in developing it. At present we can only return, but return definitely, a verdict of non-proven.[39]

To tell the sorry tale of intelligence would take us too far afield. In conjunction with the spread of IQ testing, the first decade of the twentieth century saw a dramatic shift in how "intelligence" was understood. Intelligence, the historian John Carson explains, abruptly became "understood as a differential, quantifiable, unilinear entity that determined an individual's or group's overall mental power."[40] People and races could be ordered along one linear scale, and convenient tests could identify and locate people on that scale, making the assignment of people to schools or jobs a straightforward task. It beguiles still.

To this day, scientific racists have drawn on Spearman's approach to ranking human beings and without much of what nuance he had. Remarkably they still do, in an eternal recurrence that demands ever present statistical vigilance. Each generation gets another *Bell Curve* that confidently dresses up current inequalities in the illusion of scientific rigor.[41]

From Aspiration for Data to Realization of Data

Nightingale, Galton, and Quetelet dreamed of the systematic collection of data on people to guide policymakers; Pearson organized teams of primarily women to undertake the labor of collection and analysis. These reformers' dreams of rendering people as stable bureaucratic entities came to be realized from the early twentieth century onward.[42] In the first half of the twentieth century, the collection of data on populations expanded dramatically, in the United States and Europe and in their colonial possessions. Forms for recording that we now take for granted, such as the birth certificate, made people into data in a hitherto unprecedented way. And they did so only through tremendous, consequential, contested work, what historian Wangui Muigai describes as the "interactions, confrontations, and disputes over how individual people should be accounted for and the labor involved in constructing and documenting those identities."[43] As German critics of "vulgar statistics" had recognized 150 years before, putting things into numerical or simple categories flattens reality and diversity. And doing so for all residents of a country not just classifies them but structures their understanding of themselves and their relationships with legal, medical, and educational authorities. In the United States, authorities sought carefully to locate each infant by a precise racial and sexual category, to secure the social order. A revised birth certificate form, Muigai explains, "became a key tool for policing, from birth, who counted as white and who counted as black in Virginia."[44] In colonial India concerns with locating people in administrative and caste categories likewise predominated. Along with birth certificates and the census, intelligence and personality testing boomed from World War I. All of this was an uneven process, much resisted and often ignored, often done less through fancy mathematical statistics than via conventional bureaucratic labors of documentation and standardization. Having struggled to undertake statistical inquiry into the deadly late colonial famine in Bengal, Mahalanobis noted that serious data collection requires building "up an efficient human organisation with carefully selected and trained staff. This takes time. And unless such time is allowed the results are often not only useless, but even harmful."[45] Historian Sandeep Mertia shows how Mahalanobis' efforts focused "on scale and standardization of computational work in conditions of limited resources and staff, . . . in India's large and linguistically and socio-culturally diverse geography."[46]

From the 1930s in the United States, government statisticians introduced techniques of representative sampling to make novel claims about the nation as a whole without exhaustively registering every person or farm. Emmanuel Didier argues that "the emergence of representative surveys accompanied, informed, and confirmed the birth of a new governmental interventionism" characteristic of the mid-century welfare state in the wake of the New Deal.[47] Authoritarian regimes in Germany and elsewhere at the same time plumbed ever further into the private affairs of their citizens and corporations toward ever greater total systems of surveillance.[48] Under Mao, the People's Republic of China, for their part, rejected sampling as bourgeois mathematical sophistry in favor of properly communist "exhaustive methods of periodic reports and censuses."[49] In all these diverse cases, the great labor of data collection and analysis made dramatic new understandings of the state, the population, and the economy possible—and actionable.

Alongside these massive data collection efforts came the making of infrastructures allowing for the automatic processing of data, long before digital computers, most notably Hollerith punched card machines. "Across the world," historian Adam Tooze argues, "bureaucrats were inspired to dreams of omniscience" thanks to the card machines: "For the first time it became possible to conceive of an entire nation recorded in a single database instantly accessible by means of mechanical handling equipment."[50] An early history of statistics worldwide noted, "The sociological value of the minuter statistical presentation of demographic data thus brought within reach, is not yet fully understood." Lacking the punched card machines, the authors explained, "we could never hope to lay bare all the truth we must have, if we are to cope successfully with the problems growing out of the heterogeneous commingling of races which our defective immigration laws are forcing upon us."[51] And yet, however much they flattened the texture of human existence, such rendering of human life into statistics processed on machines nevertheless could make stratification visible, both to justify and extend that stratification—or to contest it.

The Hubris of Scientific Racism

Of all vulgar modes of escaping from the consideration of the effect of social and moral influences on the human mind, the most vulgar is that of attributing the diversities of conduct and character to inherent natural differences.

—John Stuart Mill[52]

Much of the history of statistics intertwines with the long, sorry tale of attempts to prove that social hierarchies rest on innate differences between people, whether differentiated by sex, race, or class. We've been duped time and again by such claims, which have persisted to our genomic age.[53] Few scientific claims should be viewed with more suspicion than claims to innate difference that just happen to reflect our current social arrangements. History teaches that such claims demand the highest level of vigilance—about the data used, how that data is manipulated, and the inferences drawn from it. Often the lesson is simple: we know far less with certainty than many people proffering statistics and statistical inferences claim, and we need our own Du Boises to remind us.

In his review of Hoffman's statistical treatment of race, the Howard University mathematician Kelly Miller savaged the poor inferences and faulty data, and failure to consider alternate explanations: "It does not account for the facts arranged under it as satisfactorily as can be done under a different hypothesis. The author fails to consider that the discouraging facts of observation may be due to the violent upheaval of emancipation and reconstruction, and are, therefore, only temporary in their duration."[54]

Despite the explosion of data and new techniques, at the

turn of the twentieth century, no one knew how best to test such different, competing hypotheses. A new form of science needed to be created, to answer pressing questions: Which fertilizer encouraged the growth of barley? Which medicine worked most effectively? And that science was first created at the Guinness Brewery, in the hands of someone known to history as "Student."

CHAPTER 5

Data's Mathematical Baptism

So enormous was the excitement about the initial public offering of Guinness stock in 1886, that aspiring investors broke the door of their banker, the storied firm of Barings. Selling 65 percent of his stake in the company, Edward Guinness gained six million pounds, thus giving the company a valuation of some nine million pounds—well over $300 billion in today's dollars. Speculative folly or not, Guinness suddenly was well-resourced to transform its business using the latest technology of the day: statistics. One hundred years before the data sciences promised to revolutionize business practices, the team at Guinness tried to create an industrial science of brewing. Flush with cash like the rich firms in Silicon Valley or Shenzhen today, Guinness hired talented young scientists and engineers, appointed them to the august position of "brewers," and built them new experimental facilities. Then as today, new scientific skills threatened to render older forms of expertise obsolete—and often did so. Older forms of agricultural knowledge and expertise faded in the face of new chemical and mathematical tools.

Whether examining hops, barley, or the effects of different manures, the Guinness research scientists ran into two major difficulties: they had only a small number of observations and those observations varied considerably among themselves.[1] They needed some way to gauge which differences mattered, to know which differences were significant. A significance test, if you will. One among them, William Gosset, was more mathematically inclined. And from him came the significance test. All in the interest of better—and more profitable—beer.

IN THIS CHAPTER, we look at three scientists most associated with the creation of significance and hypothesis testing: Gosset, Ronald Fisher, and Jerzy Neyman. Each came to the problems with radically different preoccupations. Gosset wanted to identify the brewing process to yield the best beer. His was an engineering task. Fisher wanted scientific knowledge. His was a scientific task. And Neyman wanted to make the best choices. Each advanced methods for deciding among hypotheses using data and the newly christened field of "mathematical statistics." They utterly differed on what *deciding* meant—what deciding added up to.

Gosset was an industrial statistician; Fisher a gentleman scientist; and Neyman a mid-century rational planner. Gosset devised statistics for the purpose of maximizing profits by maximizing taste and consistency and durability. Fisher devised statistics to create scientific knowledge of the way the world really is. Neyman sought to help make choices in the most rational manner according to the evidence at hand. Together they created a new understanding of what it means to be scientific—by testing hypotheses using statistics.

Gossett: Testing for Beer Making

In 1923, a key Guinness employee explained the point of doing scientific experimentation: "The object of testing varieties of cereals is to find out which will pay the farmer best."[2]

Experiment was great—but expensive. Today, we worry about the enormous problems of dealing with too much data. By contrast, the problem here was too little data and the real costs of doing additional experiments to get more. The author, William Gosset, was an engineer and practical mathematician turned brewer, and the answer to the question transformed the science of statistics—and of experimentation—decisively. The techniques of Galton and Pearson were fine for large sets of data. Something different was needed, the brewer Gosset argued:

> it is sometimes necessary to judge of the certainty
> of the results from a very small sample, which itself
> affords the only indication of the variability. Some
> chemical, many biological, and most agricultural
> and large scale experiments belong to this class,
> which has hitherto been almost outside the range of
> statistical enquiry.[3]

Whereas Karl Pearson focused on large amounts of biological data, Gosset was concerned with industrial applications where the amount of observed data was small.[4] Collecting lots of measurements about people was easier and cheaper than undertaking expensive sets of industrial and agricultural experiments. This led Gosset to devise techniques for assessing the confidence in inferences based on small data sets and to minimize cost in doing so. Methods of experiment needed to maximize profit by minimizing cost. "In such work as ours," he wrote to Karl Pearson, "the degree of certainty to be aimed at must depend on the pecuniary advantage to be gained by following the result of the experiment, compared with the increased cost of the new method, if any, and the cost of each experiment."[5] No universal judgement about significance could be made in absence of cost. While on vacation, Gosset bicycled to meet Pearson and learn "nearly all the methods then in use."[6]

GOSSET'S TEST

Suppose you like beer. Even more, you like to make money from brewing beer. You want to improve the yield of barley because you brew Guinness beer. You can perform different experiments with fertilizer, irrigation, varieties, and so forth, but how would you know with certainty that something you did worked? Gosset devised a test to help him assess experiments, now known as Student's t-test. Imagine you have ten fields next to each other, and you plant one of two kinds of barley in every other one. You measure the size of the subsequent crop in each field, and compare the size of each field of barley type 1 with the size of each field of barley type 2. There will almost always be some difference. But what of the difference is due to chance variations and what's due to the various varieties? We need to determine how likely the difference is simply due to normal fluctuation and not any particular cause, in this case the different variety. In Gosset's test, we compute a statistic by dividing the mean of the differences by the standard deviation over the square root of the number of data points. We can then look up on a table how probable it is that we would get data with these qualities from random variations. If it is extremely unlikely that random variations would have produced yields of plants described by this data, then we have good grounds to think that one variety of barley grows better. Later we'd come to know this as testing a hypothesis—that one form of barley grows more—against a "null" hypothesis—that both forms of barley grow about the same.

You're probably wondering what's "extremely unlikely." Is a 1/10 chance good evidence something is just a random variation? 1/20? 1/100? 1/1000000? This is a decision of the experimenter. It's a choice. It's not given by science. It's a decision about what would make us comfortable that we have confidence that one thing is the case, and another not.

For Gosset, it was about making money for his firm. He was interested in helping Guinness maximize profit, and the level of certainty we might acquire needed to attend to that concern. So Gosset provided no one rule of statistical significance.

With his studies, Gosset provided a radical new way to think about making decisions under conditions of uncertainty. He offered new mathematical techniques for choosing one course of action over another based on inconclusive evidence. His new math helps us decide whether some evidence was conclusive enough for the application in question. Gosset called this a "pecuniary" approach to questions of what constitutes adequate knowledge for action, for making business choices.

By employing and celebrating applied scientists such as Gosset, Guinness pioneered the application of scientific reasoning to the production of beer and all the industrial and agricultural facets of the process. Variation of crops increasingly gave way to standardization of varieties. Local qualities of brews gave way to standardized and usually more stable beers. The challenges of the variation within experiments with small sample sizes motivated Gosset to explore mathematical techniques. The goals were not scientific knowledge itself. Gosset and his team sought to optimize the entire brewing process using mathematical tools, to increase quality, durability, and ultimately profits.[7]

As an employee of Guinness, Gosset was required to publish his results under a pseudonym, with all references to brewing replaced with other subjects. His work is now known under the name "Student." As we will see, the story of statistics and data science is replete with authors obscuring the data and motives that provoked their inquiries and innovations.

Published in Pearson's premier journal of statistics and eugenics, *Biometrika*, Gosset's work initially garnered little interest within the broader statistical community. That community was more focused on describing the social and natural world numerically than in sorting through hypotheses about them. In time, Gosset's ideas came to reorient the sciences and the social sciences largely through the work of Fisher and then of Neyman—each dramatically reworked Gosset's work.

Like Gosset, Fisher worked extensively on applying statistics to real world problems, above all questions of agricultural productivity. Unlike Gosset, Fisher sought scientific knowledge itself. He drew on Gosset's work on small data sets to offer a revolutionary conception of scientific experiment itself.

Discussing the great figures in the development of mathematical statistics, the statistician Florence Nightingale David explained, Gosset "asked the questions and [Egon] Pearson or Fisher put them into statistical language and then Neyman came to work with the mathematics."[8]

Fisher: Testing for Truth Making

In 1925, Ronald Fisher spared no words in denouncing the usefulness of the statistics of his day: "the traditional machinery of statistical processes is wholly unsuited to the needs of practical research. Not only does it take a cannon to shoot a sparrow, but it misses the sparrow!" The cannon wasn't good for small sets of data: "The elaborate mechanism built on the theory of infinitely large samples is not accurate enough for simple laboratory data." Simple laboratory data needed new techniques, techniques like those of Gosset, grounded in better mathematics and a better grasp of scientific experiment itself. "Only by systematically tackling small sample problems on their own merits does it seem possible to apply accurate tests to practical data."[9]

From the form of hypothesis testing Fisher developed came the billions of "p values" that have dominated much scientific production from the mid-twentieth century to this day and are legally required for accepting the efficacity of many forms of medical and pharmaceutical treatments. Fisher set down a new mathematical basis for statistics as a replacement for previous understandings of what makes up science.

Like Gosset, Fisher developed his tools in a practical agricultural context, in his case the Rothamsted Experimental Station.[10] Gosset met Fisher at the station in 1922. Like Gosset, he brought a new level of mathematical sophistication to an older experimental program.

At Rothamsted, Fisher was presented with data accumulated over generations of agricultural experimentation. And he was at liberty to help guide the design of experiments at the station. "The activities at Rothamsted," his daughter and biographer wrote, "the interests and the problems of the staff, the discussions over a cup of tea, all were a great stimulus to Fisher's ingenuity and inventiveness."[11] A stream of papers with new mathematical approaches tied to applied agricultural problems quickly appeared. His papers soon focused on the "significance" of observed variations for the first time.[12] The problems could often be, well, shitty to solve. "It is not infrequently assumed," one paper began, "that varieties of cultivated plants differ not only in their suitability to different climatic and soil conditions, but in their response to different manures." The fundamental question was how to move from anecdotal evidence to more "conclusive evidence as to the relative value of different manures."[13]

The answer came from statistical testing, the sort of testing Gosset had pioneered. Drawing equally upon his skill in math and the experience of agricultural testing at the agricultural station, Fisher recast Gosset's approach into a new account of scientific experimentation itself.

In 1925, he integrated his approaches into a textbook, *Statistical Methods for Research Workers.* The book spread his approach to experimentation widely. "Daily contact with the statistical problems which present themselves to the laboratory worker has stimulated the purely mathematical researches upon which are based the methods here presented."[14] The book decisively moves from older statistics interested in "aggregate, or average, values" to the "study of the causes of variation of any variable phenomenon, from the yield of wheat to the intellect of man," which requires "the examination and measurement of the variation which presents itself."[15]

Creating an experiment involved the statement of the hypothesis to be tested against a null hypothesis. A significant result in favor of hypothesis means that we believe that we would only obtain the data from an experiment from a null hypothesis a very small percentage of the time, say one out of twenty, or 5 percent. While denying any universal threshold, Fisher argued, "it is usual and convenient for experimenters to take 5 per cent. As a standard level of significance, in the sense that they are prepared to ignore all results which fail to reach this standard."[16]

Fisher's austere doctrine was designed to ward off the many biases, the hopes, the dreams that cloud judgment in thinking about data. How to eliminate all the various potential causes, seen and unseen, that might disrupt our efforts to isolate one potential cause for investigation? To avoid the often unconscious ways scientists might tilt the scales by selecting comparisons and to rule out the countless other causes that might muck up an inquiry, Fisher insisted on the need for randomization in the creation of an experiment. Fisher insisted upon randomization of things to be tested so that "the test of significance may be guaranteed against corruption by the causes of disturbance which have not been eliminated." Randomization, Fisher explained, "relieves the experimenter from the anxiety of considering and estimating the magnitude of the innumerable causes by which his data may be disturbed."[17] To preclude the dangers inherent in manipulating data after an experiment, Fisher required that the plan for the analysis of the data and the hypothesis to be tested be locked into place before data collection began. In doing a trial of a pharmaceutical, for example, we must use, and typically preregister, some procedure to choose randomly which patients will receive the drug we're testing, and which patients will receive a placebo.

While many of Fisher's demands echo as essential to good scientific practice to this day, other of his demands still cause consternation. Critics from his time onward have challenged randomization as wasteful at best and unethical and deadly at worst. In pharmaceuticals, the gold standard of randomized controlled trials (RCTs) has unquestionably protected consumers from negative side effects and ineffective drugs, but at the cost of slowing drug approval and of narrowing the grounds for deeming therapies effective. The lag in experimental treatments coming to market, long deplored by industry, became a rallying point in the movement to recognize and treat HIV infections in the 1980s and 1990s.[18] Long before that, Gosset tried in vain to get Fisher to acknowledge the inefficiencies required for randomized trials.[19]

LIBERTY, EUGENICS, AND THE UPLIFT OF RACES

Gosset wanted a better procedure for testing ingredients to make beer. Fisher sought nothing less than to enhance human freedom through knowledge based on experiment. "The liberation of the human intellect must, however, remain incomplete so long as it is free only to work out the consequences of a prescribed body of dogmatic data, and is denied the access to unsuspected truths, which only direct observation can give."[20] Only experiential knowledge could overcome dogma. Justifying experimental procedure has long bedeviled philosophers: how can we come to generalizations from individual experiences? Writing as fascism was spreading across Europe, Fisher argued that human freedom required experiment: "the arts of experimental design and of the valid interpretation of experimental results, in so far as they can be technically perfected, must constitute the core of this claim to the exercise of full intellectual liberty."[21]

For Fisher, science was not a mechanical enterprise of improving profit. For him, human progress involved "not a question merely of producing a highly efficient industrial machine, or a paragon of the negative virtues, but of quickening all the distinctively human features, all that is best in men, all the different qualities, some obvious, some infinitely subtle, which we recognise as humanly excellent."[22] His vision was less one of universal uplift than conflict among the different races of humanity. In one youthful piece, Fisher explained, "The widespread, fruitful, and successful races of the future belong to the dominant nations of to-day; and nations are rendered dominant principally by the loyalty, enterprise and co-operative ability of the people who compose them."[23] Fisher saw a heightening of conflict among national races but did not cast human progress in industrial terms. Fisher's eugenics was a heady stew, an unlikely combination of Darwin, Nietzsche, and Anglicanism. A race war was afoot, but it was, he maintained, not an *industrial* race war.

Ghosts of the decline of great civilizations haunt Fisher's

biological work. Like other eugenicists, he sensed a striking inversion of human reproduction. In economically advanced, market-driven civilizations, the most financially and culturally successful were the least reproductively successful; the best of society, he argued, were slated not to reproduce adequately, and would be swamped by lesser human beings. The economic relations of civilized societies, from Rome to contemporary Britain, were dysgenic. And the progressive elimination of the best people led inexorably to civilizations losing the higher accoutrements of human culture and eventually to the decline of those civilizations and the races that had created it.

In other words, to permit an economic logic to dominate was to lose the best of humanity, both the highest forms of culture, including science, and the most genetically superior people. Given his eugenical framework, it is hardly surprising that Fisher reacted angrily to statistical rivals who envisioned hypotheses testing along the lines of economic efficiency. They were turning the bulwark against dogmatism into a lesser cultural form.

AGAINST COST FUNCTIONS

As we saw, Gosset evaluated the efficacy of an experiment in terms dictated by potential profit. Later versions of this pragmatic understanding horrified Fisher, who came to understand it as violating the purity of scientific inquiry with base industrial motives. Fisher explained that no pecuniary value could decide the contours of knowledge:

in inductive inference we introduce no cost functions for faulty judgements, for it is recognized in scientific research that the attainment of, or failure to attain to, a particular scientific advance this year rather than later, has consequences, both to the research programme, and to advantageous applications of scientific knowledge, which cannot be foreseen. . . . We make no attempt to evaluate these consequences, and do not assume that they are capable of evaluation in any sort of currency.[24]

Fisher's greatest rival disagreed. Rather than pining after scientific truth, statistics needed to focus on making choices—in business as well as in science.

Neyman: Testing for Decision-making

The problem with most hypothesis testing, the Polish mathematician Jerzy Neyman argued for decades, is that most people thought it was about truth. Neyman argued it was about choices. "Without hoping to know whether each separate hypothesis is true or false, we may search for rules to govern our behaviour with regard to them, in following which we insure that, in the long run of experience, we shall not be too often wrong."[25] We needed more efficient tests, not the truth: "no test based upon the theory of probability can by itself provide any valuable evidence of the truth or falsehood of that hypothesis."

Neyman and his collaborator Egon Pearson, son of Karl, argued that Fisher had failed to appreciate a second danger in hypothesis testing. Fisher worried about accepting a hypothesis that was false; Neyman and Pearson stressed the need to worry about rejecting a hypothesis that we should accept as true. In testing hypotheses, then, we need to balance two types of error, quickly branded type I and type II. Fisher advised always testing a hypothesis against a null hypothesis. Neyman and Pearson insisted on the need to compare competing hypotheses.

How did this dramatically different approach to statistical

testing arise? Neyman brought together extremely recondite mathematics, a skeptical vision of knowledge, and practical agricultural work. Statistics was not Neyman's obvious vocation. After failing spectacularly as a student of experimental physics, Neyman, a Polish student living in wartime Russia, became caught up in the effort to recast mathematics on highly abstract theoretical grounds. This mathematics seemed distant from all practical application, even theoretical physics. Jobs were few, then as now, for pure mathematicians climbing the icy slopes of abstraction, and before long Neyman found himself working in highly applied statistics jobs to pay the bills and to provide housing.

His views emerged amid the experimental agricultural work in newly—and temporarily—independent Poland in the 1920s. Neyman was working within what the science historian Theodora Dryer calls "a dynamic movement to imagine sovereign Poland as a modern and prosperous agrarian nation state" through experimentation and the most sophisticated theoretical mathematics of the day.[26] Like Fisher, Neyman and his colleagues found powerful tools for analyzing agricultural experiments in Student's papers. Neyman's approach should be understood within a dream of rational creation of a thriving economy.

How did he connect his beloved highly abstract mathematics with this concrete agricultural work? Neyman drew upon a favorite book, *Grammar of Science*, by the English eugenicist and statistician Karl Pearson.[27] There he imbibed Pearson's vision of science attacking, Neyman said, "in an uncompromising manner all sorts of authorities," throwing off all existing dogma, whether religious, social, or scientific—heady stuff for young people in Russia on the verge of overthrowing tsars and church alike. He embraced Pearson's profound skepticism about what we really know. Later in life, Neyman explained,

One of my favorite ideas, learned from Mach via Karl Pearson's "Grammar of Science", is that scientific theories are no more than models of natural phenomena, frequently inadequate models. A model is a set of invented assumptions regarding invented entities such that, if one treats these invented entities as representations of appropriate elements of the phenomena studied, the consequences of the hypotheses constituting the model are expected to agree with observations. If, in all relevant trials, the degree of conformity appears to us satisfactory, then we consider the model an adequate model.[28]

For Pearson, knowledge was always provisional. "Belief," he explains, is "to be looked upon as an adjunct to knowledge: as a guide to action where decision is needful, but the probability is not so overwhelming as to amount to knowledge."[29] The best we can do is assert confidence in one model as best conforming to the phenomena we have at hand. We don't have any real insight into the true causes and things at work around us. Neyman's task was to show how the highly abstract mathematics he adored can help evaluate and construct models.

Money was always short for the young scholar, so Neyman held a series of applied statistical positions, until he received funding to travel to England to work with Karl Pearson, his old hero. To his surprise, Pearson knew little of the new abstract mathematics, but helped Neyman secure a fellowship to Paris, where Neyman returned largely to the world of abstract mathematics.

Here Gosset returns to our story. Again he inspired the more mathematically inclined statisticians. In 1926 Gosset wrote to Pearson's son Egon with a series of questions about the meaning of hypothesis testing.

If there is any alternative hypothesis which will explain the occurrence of the sample with a more reasonable probability, say .05 . . . , you will be very much more inclined to consider that the original hypothesis is not true.[30]

And this letter inspired Egon Pearson to write Neyman in Paris, bringing him back into the statistical fold. This paragraph contains "the germ of that idea" of the radical alternative that Neyman and Pearson created to Fisher's conception of testing and of scientific knowledge.[31]

In place of truth, Neyman advocated looking for reasons to pursue one set of actions, and not another. "Deciding to affirm" something scientific, Neyman wrote, "doesn't mean knowing or even believing." Rather, "it's an act of will preceded by some experience and deductive reasoning, just as one takes out life insurance, which we do even if we expect to live for a long time."[32] The statistician Erich L. Lehmann noted the dramatic significance of this new point of view. "For the first time it states the aim of statistical theory to be *the systematic search for optimal procedures*. Much of the theory developed during the next decades was directed toward this end."[33]

Neyman's views were anathema to Fisher.[34] Fisher had tried to use the new tools of statistics to explain how inductive knowledge was possible. Neyman used them to deny the existence of such knowledge, in favor of making decisions based on evidence: at best, we don't gain inductive knowledge, he said, but rather an "inductive comportment" on the basis of evidence.[35] Fisher and Neyman (and his colleague Pearson) would fight for the next thirty years. Often arcane in appearance, their arguments revolved around the adequacy of mathematics for resolving questions of human knowledge and human behavior.

Fisher argued that Neyman misunderstood what had allowed the great developments of science during the previous two hundred years: "the continuous development of mathematical thought in Western Europe from the great French mathematicians of the 17th century onward, has come to fruition in our own time, by cross-fertilization with the Natural Sciences, in supplying just such a model of the correct use of inductive reasoning, as was supplied by Euclid for *deductive* logic."[36] But the errors were deeper. Misunderstanding science meant that Neyman and his legions of followers misunderstand how knowledge can set you free, and thereby had become allies of nothing less than totalitarianism.

> To one brought up in the free intellectual atmosphere of an earlier time there is something rather horrifying in the ideological movement represented by the doctrine that reasoning, properly speaking, cannot be applied to empirical data to lead to inferences valid in the real world. It is undeniable that the intellectual freedom that we in the West have taken for granted is now successfully denied over a great part of the earth's surface. The validity of the logical steps by which we can still dare to draw our own conclusions cannot therefore, in these days, be too clearly expounded, or too strongly affirmed.[37]

The Truth Algorithm: Cookbookery and "p-value-ology"

In the aftermath of World War II, two distinct legacies of all this statistical effort pulled in divergent directions. The first led to the dramatic upending of what it meant to do science, indeed to *be* a science, in the second half of the twentieth century. The second led toward professionalized statisticians pursuing the rigor of abstruse mathematics in the style of Neyman, often distant from everyday uses of problematic real-world data.

The fights to understand experimental results statistically, and to use data to adjudicate between competing hypotheses, had lasting impact that still shapes our world and our sensemaking today. The most visible impact is the ubiquity of statistical significance in our thinking about "chance" events, with the algorithmic understanding of a result being true if we can establish a probability—known colloquially as the "p value"—below the magic number of .05.

To be clear, such an algorithmic approach to setting truth would have been anathema to Fisher and to Neyman and Pearson alike.[38] The force of their arguments, however, was unable to outlive them in the face of the growing demand for objective certainty and for rational decision making. Over the course of the second half of the twentieth century, searching for effects improbable under a null model became the criterion for publication, for approval of drugs, and in more popular discussion, for separating chance from causation. When Fisher first published his dramatic new account of experimental design, critics abounded. In time, through the agency of easier-to-read textbooks, hypothesis testing became central to a wide array of sciences.[39] Science historian Christopher Phillips has explained, "food scientists, psychologists, sociologists, and physicians . . . saw statistical methods as providing an off-the-shelf technique to make reliable causal judgments in inescapably subjective settings."[40] While Fisher himself inveighed against a cookbook approach to experiment, the suggestion that .05 was a good threshold for determining significance became essentially the line between a publishable result and scientific garbage in the second half of the twentieth century. It provided a false sense of objectivity.

In field after field, these new quantitative approaches dis-

rupted older visions of who was an expert and what expertise in a field meant. Nowhere was the shift so dramatic and
significant as in the investigation of the efficacy of pharmaceuticals. The randomized trial upended the authority of
physicians in judging that efficacy.

In 1961, the American Medical Association denied that
anyone other than practicing physicians should opine on the
utility of a treatment: "the only possible final determination
as to the efficacy and ultimate use of a drug is the extensive clinical use of that drug by large numbers of the medical profession over a long period of time."[41] Physicians and
pharmacists resisted the loss of their control over questions
about the efficacy of drugs through the middle of the twentieth century, before dramatically losing ground in 1962, when
the Food and Drug Administration (FDA) acquired dramatic
new powers. Regulators, Theodore Porter explains, "considered that the expertise of doctors provided an inadequate
control on the bold claims of drug manufacturers. The alternative was a more centralized decision process, to be based
mainly on written information."[42] With the 1962 law, called
the Kefauver-Harris Amendment, the randomized controlled
trial became the benchmark for gauging the efficacy of med-
ications, to become the gold standard for authorization of
drugs and the documentation of their side effects. The legislation enabled the FDA to gauge drugs going forward and to
look retrospectively at medicines already on the market and
to remove from sale dangerous or useless drugs previously
approved between 1938 and 1962.

And what of the debates of Fisher and Neyman/Pearson?
Few cared about philosophical niceties in applying hypothesis testing. Abstruse philosophical debates tend to disappear
when things are given wide application. And so it was with
hypothesis testing. "Fisher's theory of significance testing . . .
was merged with concepts from the Neyman-Pearson theory

and taught as 'statistics' *per se* . . . it goes without saying that neither Fisher nor Neyman and Pearson would have looked with favor on this offspring of their forced marriage."[43]

Mathematics, Not Data: The Placement of Statistics after World War II

With the entry of the United States into World War II came an explosion of statistics applied to war work, centered at Columbia in New York, Princeton in New Jersey, and Berkeley in California. "The only useful function of a statistician," wrote W. Edwards Deming of the US Census Bureau, "is to make predictions, and thus to provide a basis for action."[44] Ranging from how best to set proximity fuses, to quality control in factories, to the best angles for torpedoes, the wartime successes of these statistics groups were legion. Highly applied statistics spurred the creation of new approaches to the analysis of experiments, above all a form of testing called sequential analysis. Sequential analysis unified quality control procedures from Bell Telephone Laboratories with the economistic approach to testing like that Gosset celebrated.[45]

Before the war, a few statisticians had attempted to recast their approaches in the idioms and procedures of pure mathematics and agitated for the cordoning off of mathematical statistics from its more data-driven applied cousins. Ironically, wartime successes with applied statistics came to justify this move toward abstraction. The "needs of the war," a 1946 document from the Office of Naval Research explained, "gave impetus to basic research which resulted in the formation, at Columbia University, of the new theory of Sequential Analysis."[46] Key figures such as the Columbia statistician Harold Hotelling leveraged all these successes to justify the support for highly theoretical and highly mathematical statistics; a remarkable mathematician and program administrator Mina Rees supported them from within government. The stories of their wartime successes explained that theory made their successes possible. Immediately after the war, at a moment when it seemed that science might no longer receive major government funds, the Office of Naval Research accepted this story. "That progress in mathematics is basic to progress in science is generally recognized; but it was forcefully demonstrated during World War II."[47] In doing so, it freed ample military funds for extremely theoretical statistics.

As a result, the highly data-focused work of the war shifted in emphasis. "During World War II," Neyman wrote in the late 1940s, "the majority of statisticians were working on problems of defense which frequently bore the imprint of immediate practical importance." He held a major symposium "to stimulate the return to theoretical research."[48] Hotelling explained how applications beckoned—but also corrupted—statisticians: "the call of application is enticing, and has led many young scholars to forsake the cultivation of statistical theory."[49] Abstraction was all the rage in mathematics, and it appealed likewise to statisticians working in Neyman's mode.

Neyman's career at Berkeley exhibited tremendous advances in mathematical statistics, but also in defining mathematical statistics as a discipline. To be clear, this was not entirely an academic quest: Neyman also wanted Berkeley to recognize that his group should be a full department, not merely a "lab" within the department of mathematics. Doing so required Neyman to establish the mathematical bona fides of mathematical statistics—to show that the field was sufficiently rigorous, intellectually and mathematically, to warrant a department.

In retrospect, it's not clear that the relationship between data and our daily lives hinges on how mathy is the underlying analysis. At the time, however, mathematical statistics as a profession aligned itself ever more with the rigor and the axiomatic approach associated with pure mathematics. Like the Office of Naval Research, the new National Science Foundation (NSF) accepted the view that theoretical, mathematical statistics had made the wartime applied successes possible, and funded statistics accordingly. Since the founding of NSF in 1951, statistics has been located as mathematics: not as an aspect of engineering, as one might conclude from postwar activities or the economic impact in the twenty-first century; nor in the "natural sciences" as Fisher would likely have preferred. The funding, therefore—the academic lifeblood of an intellectual field—has relied largely on establishing sufficient mathiness, driving the field toward what the mathematical probabilist Leo Breiman critiqued as "over-mathematization" in 2004. By 1962, the Princeton topologist-turned-statistician John Tukey argued that "data analysts" ought to "use mathematical argument and mathematical results as bases for judgment rather than as bases for proof or stamps of validity"—a strong sign of how far mathematics had permeated—or warped—statistics.[50] As we will see in the following chapters, it's an entertaining counterfactual history to imagine: Would pattern discovery have been born in electrical engineering, or machine learning in computer science, or data science in industry, had American academic statistics not become quite so mathematical from WWII until the end of the century?

PART II

CHAPTER 6

Data at War

I n the early 1960s, the cryptographer Juanita Moody regretted that her employer, the hypersecretive National Security Agency, could not put its massive capacity for analyzing data to good use in the nonclassified world: "it always worried me that we had great computerized capability just moving faster than you could imagine and that there was this whole, big medical world out there that needed it." As soon as she could leave the NSA, she said, "I'm going to go volunteer to do something to help the medical world with computerized data processing. You just knew that was a problem, and yet everything we were doing was classified." Worse yet, it didn't have to be that way: "And I knew it didn't have to be classified, but it was."[1] Decades before the rise of big data in the 1990s and 2000s, the National Security Agency had institutionalized data collection, algorithms, and forms of analysis.

How did this come to pass?

We need to relocate to sixty-six miles northeast of East Ilsley, where Gosset and Karl Pearson had met in 1905. There lay the quiet English town of Bletchley Park, one of the most secret and most significant sites in World War II.

Bletchley Park

While the statisticians Fisher and Neyman battled over the truths and errors, a group of outsiders was creating a radically different future combining computation, labor, and data in the context of war. The outsiders to statistics were the engineers, linguists, and mathematicians of Bletchley Park, nestled secretly between Oxford and Cambridge in England, breaking German codes. Protected by a cover story of "Captain Ridley's Shooting Party," these scientists and humanists, many recruited through old-boy networks, were in fact pioneers in building specialized computing hardware for making sense of streams of data at a very large scale.[2]

The work involved a noisy cacophony of machines, paper tape, men and women, with only a few working at desks scribbling more letters than formulas. Most of the men doing the scribbling were drawn from a variety of academic pursuits and skill in games such as chess or crossword puzzles, not from academic statistics. The most famous of them, Alan Turing, who took a train to Bletchley the day after the United Kingdom declared war in September of 1939, was a mathematician known primarily for work in logic, who had earlier worked some with statistics.

Rather than focusing on quantifying the qualities of a state's people or investigating scientific hypotheses, Turing and his colleagues worked on an applied, martial task, involving data and the world's largest (at the time) computation.

This was the *practice* of statistics and data in the most aggressive form of its era, and marks a watershed moment in our broader history: when data leapt to a pragmatic new existence defined by engineering and problem-solving.

Unschooled in the raging debates about rigor in mathematical statistics, the researchers of Bletchley Park developed special-purpose computing hardware along with their own statistical methods for breaking the "unbreakable" German cyphers (most famously the Enigma machine) employed during World War II. In fact, "the Germans were well aware of the way the Enigma could be broken. But they had concluded that it would take a whole building full of equipment to do it" according to mathematician, US naval captain, and later NSA researcher Howard Campaigne. "And that's what we had. A building full of equipment."[3] Decoding the German codes instead required computational enumeration of an astronomical number of hypotheses, each corresponding to that day's possible secret settings of the Germans' encrypting machines. Each day, additional data would refine each hypothesis's probability, with initial settings based on guesses and heuristics about the typical language employed by the German military. Mathematical rigor, the bedrock of the new academic statistics, was beside the point. Faced with a life-or-death job to be done, Turing and colleagues used what would now be called "Bayesian" methods. They deployed diverse methods using special-purpose electromechanical computing devices, called "bombes," whirling noisily until a "stop," when the machines would halt to reveal a potential solution.

For all the work of geniuses like Turing, Bletchley mattered because it made data analysis *industrial*. "Bletchley Park in 1944 was not the hutted, collegiate, informal organisation of popular myth," writes the historian David Kenyon. After 1943, the "task was not to provide a fertile habitat for individual genius, but rather to scale up and industrialise the techniques developed by the master codebreakers, and to create systems allowing their methods to be applied to thousands of items of data, at speed, by staff without an Oxbridge level of education."[4]

The Bletchley effort culminated in the creation of what some historians consider the world's first "computers" in the contemporary sense of the word: digital, electronic, and programmable machines, called the Colossus. One staff member gave a sense of the volume of data and the nature of the machines: "the sprocket holes went past an electronic eye. They went past at five thousand per second, so that five thousand letters were registered per second."[5] Managing this data, registered primarily on finicky tapes, required labor, predominantly the labor of women. Work at Bletchley had a decidedly gendered hierarchy: "all of the cryptographers working with Colossus were men, and all of the operators were women," even though many of them had received a university education.[6] The military initially subjected the women operators but not the male mathematicians to drills and marches. These "demands placed on the Colossus operators reveal their superiors' unspoken presumption that women's work is by nature mundane and does not require one's full energies," science historian Janet Abbate explains.[7] Getting data in and out of the machines was hard work. Tape flew through the Colossus at forty feet per second. One operator, Eleanor Ireland, explained it "was a tricky operation, getting the tape to the right tension . . . [we] were terrified of the tape breaking."[8]

Time mattered in war. And the women like Ireland who staffed the machines pledged silence, which they kept until nearly the end of their lives, when the British government finally declassified aspects of the Bletchley effort. "My great sadness," reminisced Colossus operator Catherine Caughey, "is that my beloved husband died in 1975 without knowing what I did in the war."[9] So complete was this secrecy that the history of information technologies neglected both the devices and the teams of women, historian Mar Hicks explains, "paradoxically ensuring that British accomplishments went down in history as also-rans in a US-centric story of early electronic computing."[10]

Meanwhile, in the States

Across the Atlantic, the US Navy and Army built up increasingly large factory-size installations to process captured Axis communications using machines new and old, ranging from microfilm to IBM card-processing machines called tabulators. Despite long-standing secrecy, the United States and United Kingdom slowly developed a close cryptological relationship. When American cryptologists visited Bletchley early in 1941, they were not initially told about the machines for breaking Enigma—only generalities about the approach.[11] Relations warmed quickly: by 1942, American cryptographer Solomon Kullback visited for months, saying that, during his visit to Bletchley, the British "showed me everything, the details of their operations on the German systems. . . . They showed me the bombes and how they operated."[12] Turing himself traveled to the US, visiting the factory making code-breaking bombes as well as stopping at Bell Labs in New York City. The approaches of Turing and company soon were integrated into the cryptological workflow and machines modified to accommodate the new industrial-scale statistical analysis.

"Cryptography rearranges power," writes Phillip Rogaway. It "configures who can do what, from what."[13] World War II demonstrated this: breaking codes decisively altered worldwide power relations, by helping Allied forces to secure decisive victories through better intelligence in Europe as well as in the Pacific. In 1942, US Navy cryptographers broke encrypted Japanese naval communications that allowed them to predict the timing and nature of an attack on Midway Atoll in the Pacific. While still enfeebled by the attack on Pearl Harbor, the US Navy was able to surprise attack the Japanese fleet and secure a victory that helped give the US time to rebuild its depleted navy. The next April, Navy cryptanalysis revealed details of an upcoming tour by Japanese marshal admiral Iso-

Code Breaking at Bletchley Park, 1943. Bletchley Park Trust.
Getty Images. Identified as Colossus operated by Dorothy Du
Boisson (left) and an unidentified Wren. Identification draw from
Janet Abbate, *Recoding Gender: Women's Changing Participation
in Computing* (MIT Press, 2012), p. 15.

roku Yamamoto, making possible "Operation Vengeance," in
which he was killed, boosting the morale of US forces and
costing the Japanese Navy a leader considered "the outstand-
ing naval officer."[14] When the Allies landed in Normandy in
1944, Bletchley Park analysts and their American counter-
parts provided an unprecedented level of understanding of
the German positions in France and the Low Countries.[15]
Industrial data-crunching altered power relations dramatically. Just after the war, and years before NATO emerged,
the US and the UK cemented their cryptographic alliance and
unprecedented intelligence sharing from World War II into
a tight code-breaking relationship that persists to this day—
a long-standing formal alliance between the Anglophone

nations that quickly encompassed Canada, Australia, and New Zealand to make up the "Five Eyes."

This was not data in search of latent truths about humanity or nature. This was not data from small experiments, recorded in small notebooks. This was data motivated by a pressing need—to provide answers in short order that could spur action and save lives. Answers that could come only from industrial-scale data analysis.

The savvy heuristics expediting the computational search over an astronomical abundance of different hypotheses, and the dynamic updating of probabilities initialized by prior belief, would have been anathema to Fisher and Neyman alike, but set into motion the birth of applied computational statistics now at the heart of corporate data mining and artificial intelligence. Central to analysis at Bletchley was a form of statistics despised by the mathematicians but embraced and made industrial during the war: the Bayesian kind.

Bayes: From the Deity to Decryption

Fisher complained that Neyman and Pearson were mere mathematicians and not scientists; they retorted that Fisher was a mere scientist and no mathematician. The deepest insult of all lobbied in their math battle, however, was the accusation of being "Bayesian." And yet, Bayesian statistics turned out to be an excellent and straightforward approach for making decisions daily at Bletchley Park in evaluating possible decryptions of the enemy messages.

To illustrate the idea behind Bayesian methods: Consider a college in the midst of a COVID outbreak. Imagine that, first, each student has been tested with a "perfect" test, and all positives and negative cases are known. However, a flagrant system error results in the records being lost before students can be informed. The only surviving statistic is that 1 percent of the students have the disease. The college quickly gives everyone a rapid but less reliable test, one in which there's a 99 percent probability that the score is "positive" for sick people, and 99 percent chance that the score is "negative" for healthy people. All students testing positive are quarantined in one dorm. Imagine you meet a student in the dorm and you must decide: What is the probability this student is actually sick? The answer is straightforward when you're a Bayesian.

As a practical matter, being a "Bayesian" simply means using Bayes' rule, a definitional mathematical equation connecting things we know to a thing we want to know. What could be controversial about such a mathematical rule? Why is "Bayes" anathema? Bayesian *statistics* refers not just to using the equation but to an interpretation of probability often derided as "the subjective school." Within the Bayesian interpretation, the probability of something occurring is a degree of belief that something will occur. It is "subjective" because it is about a human being—a "subject" having the belief. In contrast, until recently, the vast majority of mathematical statisticians preferred working with probabilities interpreted as a statement of the objective frequency that something will happen in a hypothetical infinite number of repetitions of the same experiment: for example, a fair die will roll a 5 one-sixth of the time. Mathematics and science epitomized objectivity—indeed subjective was just the quality of thinking that Fisher and Neyman deplored in their different ways. All this is crucial, as part of data's most powerful impact is as a rhetorical tool invoking objective truth. But being philosophical about Bayes (a matter of interpretation) is very different from using the formula known as "Bayes' rule" (a matter of arithmetic).

Let's get back to the COVID diagnostic example: We start with "What is the probability that one would get a positive test for COVID, *given* that one actually is actually *infected*?" What we really want requires us to turn this around: "What is the probability that one is infected *given* that one has gotten a *positive test*?"[16] Written in English, this appears to be a small game of wordplay, but in fact, the slight change involves a different accounting of how we decide what is true and how we make decisions.

What's exciting about Bayes is that we can imagine coming to decisions about belief by simply doing accounting of various probabilities. This means merely tabulating carefully each of the separate probabilities. What's the rub? The problem comes in this statement, "It is known that one in one hundred people has the disease." Rarely does life hand one such a known quantity (recall that in the contrived example it was only the combination of a perfect test with a flagrant system error that allowed us to do so). This probability can only be calculated if we know or can estimate the overall probability that someone has the disease irrespective of the test result!

This problem emerged in the initial articulation of the rule, in a posthumously published essay by one Thomas Bayes, an eighteenth-century minister and scholar. This essay contained the crucial insight that the probability of data and hypothesis must be the product of two terms: the probability of the data given the hypothesis, and the probability of the hypothesis itself. It is the second that is the challenging quantity just mentioned. It's often called Bayesian "prior": it's a funny beast, as it is "prior to" and independent of experimental data—and in principle computable even prior to performing an experiment. What does it mean to know if a hypothesis is probable in the absence of all experimental and observational data?

While Bayes does not mention God in the original essay, the historian Stephen Stigler argues that Bayes wrote it to refute an argument by the Scottish philosopher David Hume regarding the likelihood of Christ's resurrection, a use of Bayes' rule which has continued to the present century.[17] The notorious skeptic Hume wished to compute the probability of the resurrection given the existence of reports of miracles. Mathematically, Bayes' rule states that the probability of the miracles being true given that they've been reported is the probability that they would be reported given that they were true times the prior probability that there are miracles, divided by the probability that miracles would be reported (whether or not they're actually true). An equivalent but slightly more interpretable question is: What are the odds? That is:

$$\frac{P(\text{real miracles given reported miracles})}{P(\text{no miracles given reported miracles})}$$

$$= \frac{P(\text{reported miracles given real miracles}) \times P(\text{real miracles})}{P(\text{reported miracles given no miracles}) \times P(\text{no miracles})}$$

Here the problem becomes very clear. Even if we have consensus as to the probability someone would report a miracle given that it really happened, we may have greatly differing opinions as to the prior probability that miracles take place— that is, *P(there are real miracles)*.[18] And without agreeing on the numerical value of these prior probabilities, we cannot agree on the probability that miracles took place given that they were so reported, even if we agree on the likelihood they would be reported whether or not they occurred. The reliance of a number so important on another number so subjective has remained the crux of a long-standing anti-Bayesian position among mathematical statisticians. For example, one would need to have such prior probability of divine existence (in the case of Hume) or of two competing scientific hypotheses (in the case considered by Neyman).

Despite all these serious objections, just such Bayesian analysis—at industrial scale—was at the heart of the efforts of Alan Turing and the codebreakers of Bletchley Park. In an introductory treatise he wrote at Bletchley Park, Turing explained, "Nearly all applications of probability to cryptography depend on the factor principle (or Bayes' theorem)."[19] They put just this type of decision, framed in terms of Bayes' rule, to work at the dawn of digital computation during World War II.[20] The analogous equation in the case of the sick students would be:

$$\frac{P(\text{sick given positive})}{P(\text{healthy given positive})}$$

$$= \frac{P(\text{sick}) \times P(\text{positive given sick})}{P(\text{healthy}) \times P(\text{positive given healthy})}$$

$$= .01 \times .99/(.99 \times .01)$$
$$= 1 \text{ (i.e., even odds)}.$$

That should inform your decision, for example, as to whether you should stay out of that dorm.

In place of certain priors, codebreakers at Bletchley had to rely on heuristics and guesses, such as the frequency of letter usage in German. Why were codebreakers willing to follow the types of accounting listed above? One justification for its use is that, in the limit of large data sets, the likelihood for one hypothesis so outweighs that of the competitor that the decision depends only weakly on the unknown priors themselves. An NSA paper explicitly noted, "there can exist for the cryptographer no assignment of a priori odds (whether ingenious or otherwise) that can adversely affect the usefulness of our computer program."[21] What was then a successful innovation for computational cryptography is now commonplace in data-driven applications. With great popularity, Bayesian approaches are now a signifier of statistical sophistication, rather than something to be ashamed of![22]

While these techniques remained classified secrets for decades, the new computational approaches and attitudes migrated slowly outside of the intelligence world. Turing's work was shared with only a few collaborators and allies, yet the approach left a deep impact on both sides of the Atlantic. In 1942, Turing made an extended and dangerous trip, given German U-boats patrolling the Atlantic, to Bell Labs to discuss cryptography with Claude Shannon, John Tukey, and other future luminaries of American applied computational statistics. Turing's close collaborators I. J. Good and Donald Michie went on to become leaders in the new fields of computational statistics and "machine intelligence" over the next fifty years. For decades after the war, Good served as one of the most persistent and convincing proselytizers for the use of Bayes in statistics more generally. Spending much of his career at Virginia Tech, Good continued to collaborate closely with the US National Security Agency and its British counterpart GCHQ (Government Communications Headquarters) in still-classified capacities while spreading the Bayesian gospel. In a torrent of well-written papers and books, Good spelled out interesting statistical problems best framed using Bayesian inference, often with cryptic comments that the method first was proposed by Alan Turing.[23] Throughout the Cold War, the NSA and GCHQ pursued a rich program in computational statistics following these developments, in ways that remain mostly classified.

After World War II

Elements of this thriving practice existed as a subculture outside the world of intelligence until the 1980s, when the availability of computing resources, even on microcomputers, enabled a vast academic and commercial expansion of computational statistics. As within the intelligence services, the philosophical debates around mathematical rigor or the proper interpretation of probabilities took on a character quite unlike that in academic circles.

Math mattered. But so did engineering for data. Data analysis needed its Turings, to be sure. But it equally needed its engineers, its operators.

As important as new mathematics for analyzing data was the engineering to store and to process it. On one sad day in 1948, called "Black Friday," Soviet encryption abruptly became largely impervious to decryption by the United States and its allies. These codes shifted the computational needs of the National Security Agency, founded in 1952, toward the processing of ever more vast stores of data. By 1955, more than two thousand listening positions produced thirty-seven tons of intercepted communication that needed processing per month, along with 30 million words of teletype communications. China alone produced some 250,000 intercepted messages.[24] The NSA needed the capacity to process large amounts of data far more than the capacity to perform arithmetic quickly. The data processing needs were extraordinary and outpaced the technology of the time. Historian Colin Burke explains that in the middle of the 1950s, "NSA became entangled in one of the great techno-gambles in American history: it shunted tens of millions of dollars to computer companies" to help overcome the Soviet encryption system.[25]

IBM scientist Frances Allen, the first woman to win the

Turing Award, described what NSA needed in a machine: "a streaming machine, which could take information that was gathered from the listening stations that NSA had around the world—mostly listening to Russia at the time, the Soviet Union—and then take that vast amount of data, some coded, some open, and do code breaking on it."[26] Large data meant large machines: "attached to this machine was a [tractor] tape system, which contained vast amounts of information, and the information could stream from the tape system through the Stretch Harvest memory, through the decoding unit, the Harvest unit, and then back out—the answers, whatever the results, back out without ever stopping."[27] She later explained, "It was a great giant cartridge system where the tapes had addresses and were automatically programmed to pull up a tape, bring it up to a reader and then take it off, then read it."[28] The machine was a giant pattern recognition device working in real time on this data stream and needed a programming language optimized for this purpose.[29]

In focusing "on the manipulation of large volumes of data and great flexibility and variety in non-numerical logical processes," the NSA had needs more akin to large businesses than to physicists.[30] Just as substantial federal funds promoted the creation of ever faster arithmetical machines, substantial federal funds for cryptography sponsored intense work on larger storage mechanisms. The two came together, with great friction, in funding IBM's attempts to create a jump in capability in the mid-1950s.

If cryptologists had made the very first computers, one NSA pioneer mused, their name might have been "analyzers" or "information handlers" or, even, "datalyzers."[31] Under the sponsorship of the US national laboratories concerned with nuclear weapons, computer developments focused to a great extent upon improving the processing speed needed for simulating explosions. They needed lots and lots of multiplication, not large-scale data analysis.*

The NSA funded IBM and Remington, just as they would later heavily fund Control Data Corporation (CDC) and Cray, to create computers that could perform arithmetic faster, but—perhaps even more importantly—contend with more data, often in parallel and in real time.[32] From the 1970s onward, NSA would lose much of its control over the future design of supercomputers, but it remains a primary—if not the primary—market for such machines.

Data as Engineering

Communities within the NSA approached the data with statistics much like the figures at Bletchley Park: as an engineering problem more than a scientific one.[†]

---

* The Atomic Energy Commission's "computer requirement emphasized high-speed multiplication, whereas the NSA's emphasis was on manipulation of large volumes of data and great flexibility and variety in non-numerical logical processes." Samuel S. Snyder, "Computer Advances Pioneered by Cryptologic Organizations," *Annals of the History of Computing* 2, no. 1 (1980): 66.

† "We find ourselves today in the position of beginning to be a factory. To some it is not as much fun when you don't see the data. I think one of the biggest developments was when the target countries began to use teletype equipment and began to send their data electrically. We thought at one time that we would have a mile and a half of cards, and that we would have the whole building filled with key punch operators to punch all the data; but fortunately the target countries began to be our key punch operators, which led to our being able to forward this data electrically. We are currently handling by electrical circuits some [redacted] per day which come directly into the building and are handled automatically. . . . Much of this data is never seen by any particular person. In some cases, the results go back within less than a minute, having really never been seen by an individual. That doesn't mean that much analytic work doesn't go into the preparation of the data." Joseph Eachus

They needed different computers. And they needed different math, math in the tradition of the work at Bletchley Park. While NSA mathematics remains highly classified, a small number of declassified works show that the agency pursued not simply computational statistics, but large-scale computational statistics on data being received in real time. The NSA had personnel with a tremendous degree of mathematical sophistication combined with ever-increasing streams of data and computers custom-built to contend with data; unlike academic statisticians, however, they did not have to work to justify themselves as mathematicians. Efficiency is key: the cost of large-scale computation figures centrally in the declassified NSA papers, even with all the juicy bits redacted. In a Bayesian paper on judging large numbers of hypotheses about the proper decryption of a message, the analysis produced a costly estimator needed in the analysis: "This would cost almost as much as doing the actual testing of the hypotheses. Hence, from a COMSEC [communications security] point of view, the above expression" for it "is not practically useful."[33] A communications security point of view requires not the purity demanded by academics, but a balance between statistical rigor and the requirements of vast data. "In cryptanalysis," another paper notes, "we frequently perform a million or more consecutive experiments, with a Bayes Factor computed for each experiment." Indeed, papers in the journal explicitly reject the concerns of statisticians and philosophers with the use of Bayes in the absence of a priori probabilities.[34] Given the mission, values other than the philosophical and statistical ones like we saw in

et al., "Growing Up with Computers at NSA (Top Secret Umbra)," *NSA Technical Journal* Special Issue (1972): 14.

Fisher and Neyman must prevail. Bayesian analysis was too powerful at great scale.[35]

Before large-scale algorithmic models drawing upon the automatically accumulated data of everyday business transactions upended media and advertising from the 1990s onward, the NSA had internally developed its form of computational heavy statistical machine learning focused on high volumes of messy data generated in real time. Like future machine learning it drew heavily but selectively upon statistics, and like contemporary data analysis the agency wrestled with the demands of practical databases, but with different ends.

However classified all this work, computational attitudes toward data and transformations in storage technologies slowly found their way into the nonclassified world, perhaps most famously a statistical distance named after two NSA scientists, Kullback-Leibler divergence.

The sciences followed suit. In 1950, Mina Rees of the Naval Research Office noted the "great emphasis" on early machines "that would accept a small amount of information, perform very rapidly extensive operations on this information, and turn out a small amount of information as its answer." Now, she wrote, the interest "seems to lie in a further exploration of the use of machines to accept large amounts of data, perform very simple operations upon them, and print out, possibly, very large numbers of results."[36] The experimental data produced in high-energy physics quickly challenged storage and processing abilities alike.[37] In science as in snooping, the data potentially to be analyzed and stored has ever outstripped processing power, memory, and storage capacity. "Over the past 40 years or more," a piece in *Science* noted in 2009, "Moore's Law has enabled transistors on silicon chips to get smaller and processors to get faster. At the same time, technology improvements for disks for storage cannot keep up with the ever-increasing flood of scientific data generated by the faster computers."[38]

Perhaps nowhere did this intellectual thread flourish in the wake of World War II as at AT&T's Bell Labs, where the data was not about codes and ciphers but about communications more generally: phone calls across the United States and abroad.

Data at the Labs

We [NSA] had very close contacts with the Bell Laboratories.
They were very, let's say, willing to work along with us.
—*Solomon Kullback (1907–1994), who spent 1942 at Bletchley before a distinguished career as chief scientist at NSA, interviewed in 1982*

Like Bletchley and the NSA, Bell Labs was an early example of computing with data. And Bell's data focused on people and their communications—decades before those became the lifeblood of the internet.

The Google Research of its day, AT&T's Bell Labs worked directly with data and information about people within a government-tolerated monopoly with all the data, all the researchers, and all the computing power. While they maintained close ties with academia, Bell researchers emphasized their work's distinctness from the academic traditions and shibboleths.

In a 1962 manifesto, the Princeton–Bell Labs mathematician John Tukey called for a new approach he dubbed "data analysis" that would be more dedicated to discovery than to confirmation through mathematical proof. As a scientific practice, Tukey argued, data analysis is an art, a form of judgment, not a logically closed discipline, and he encouraged the creation of new tools, from graph paper to computer graphics, to enable discovery.

Spies pioneered large-scale data storage, as its necessity became apparent within the American intelligence community soon after the Second World War. The business world soon began catching up. Starting with the data from airline reservations systems in the 1960s, industry began accumulating data about customers at a rapidly accelerating rate. In the subsequent twenty years, corporations collected data of everyday transactions: credit card purchases at particular locations, airline trips, car rentals, and later, checkouts at libraries. Over the decades of development of computers for business purposes, they were adopted by a variety of other companies such as IBM who sought new ways to turn data—primarily data about consumers—into profit. By the mid-1970s, a growing number of libertarians, government officials, and spokespersons for consumer safety took notice. "We are coming to recognize," the head of the Rockefeller Foundation noted, "that organized knowledge puts an immense amount of power in the hands of people who take the trouble to master it."[39]

Despite the wartime experience with data and its powers, and the industrial flourishing of computing with data thereafter, hopes for the new digital computers among academics and mathematicians in the 1940s focused on them as logical machines—not data processors. Just as statisticians gravitated toward abstract math, most of the early proponents of intelligent machines that exploded from 1950 onward focused on logic and math, not data about people and things.

CHAPTER 7

Intelligence without Data

Dreaming of Learning Machines

"My fondest dream," the Bell Labs scientist Claude Shannon wrote a former teacher in 1952, "is to someday build a machine that really thinks, learns, communicates with humans and manipulates its environment in a fairly sophisticated way."[1] In the wake of World War II, engineers, mathematicians, sociologists, and neurologists all speculated: Might machines perform tasks previously viewed as exclusively the province of human intelligence? A key question was, Whose intelligence? Mathematicians? Linguists? Calculators? Expert bakers? Of all the forms of intelligence in the world, the answer most typically given in the years after World War II was intelligence of the sort that people like the researchers prioritized: proving theorems, playing chess, navigating bureaucratic systems efficiently.

You might expect the analysis of data to be central to this project. It wasn't. Today, artificial intelligence primarily means machine learning on huge data sets. It didn't then.[2]

Turing

In 1950 Alan Turing published an epochal paper defending the possibility of machinery performing a range of activities typically thought to require intelligence. He rebutted arguments that computers couldn't be original, that they could only follow rules without adapting, that they couldn't learn from experience in the world. Famed for his results in logic, Turing was not unduly celebratory of logic as the pinnacle of human intelligence. His views were far more ecumenical, covering a wide range of creative, intelligent, even emotional activity.

Before his time in Bletchley, Alan Turing had published a critical result in the history of mathematics and logic, several years before any digital computer. He introduced the idea of an abstract universal machine (now referred to as a "Turing machine") that could perform nearly any logical operation. During the war, he and others at Bletchley Park knew as much as anyone about drawing tentative conclusions from masses of data. They spent their evenings speculating on the possibility of machines acting intelligently. After the war, Turing envisioned a variety of machines that might be capable of doing apparently intelligent acts drawing upon logic and data alike.[3]

In his paper "Computing Machinery and Intelligence," Turing converted a parlor game of guessing whether a hidden person was a man or woman into an operational approach to discerning whether a machine exhibited intelligent behavior, the imitation game: "The object of the game for the interrogator is to determine which" of two hidden people called A and B "is the man and which is the woman." Turing suggests replacing the Man A with a machine:

> We now ask the question, "What will happen when a machine takes the part of A in this game?" Will the interrogator decide wrongly as often when the

game is played like this as he does when the game is played between a man and a woman? These questions replace our original, "Can machines think?"

Rather than gauging whether a machine is thinking by asking if it thinks as human beings do, Turing asks instead for us to examine its behavior. "If . . . a machine can be constructed to play the imitation game satisfactorily, we need not be troubled" by the objection that machines don't operate in the ways that human beings do, and thus can't be thought of as thinking in a meaningful sense.

In his paper, Turing mused expansively about machine intelligence. Despite his prominence as a logician, he gave experience and data a central role, and did not consider only formal reasoning like mathematics or games like chess. He even included activities typically not thought of as machinelike:

Be kind, resourceful, beautiful, friendly, have initiative, have a sense of humour, tell right from wrong, make mistakes, fall in love, enjoy strawberries and cream, make someone fall in love with it, learn from experience, use words properly, be the subject of its own thought, have as much diversity of behaviour as a man, do something really new.

Our incredulity that machines might do all these things rests, he argued, on our experience of limited, ugly machines, all made for a "special purpose." From this limited experience, we mistakenly conclude that machines could do none of these things. The real thing limiting existing computers, he argues, is computer memory, which he calls storage—"the very small storage capacity of most machines."[4] A large memory would enable computers to exhibit many different behaviors: "The criticism that a machine cannot have much diversity of behaviour is just a way of saying that it cannot have much storage capacity."[5] And fundamental to producing all these results would be the ability of machines to modify themselves: "a machine undoubtedly can be its own subject matter. It may be used to help in making up its own programmes, or to predict the effect of alterations in its own structure. By observing the results of its own behaviour it can modify its own programmes so as to achieve some purpose more effectively. These are possibilities of the near future, rather than Utopian dreams."[6] Turing's untimely death by his own hand in the wake of his chemical castration by the British state foreclosed finding out where his capacious vision might have led.

In his vision of intelligent machines, Turing wedded learning from data in large memory stores to computers reprogramming themselves. This was heady stuff. The anthropologist Lucy Suchman argues that efforts in artificial intelligence function "as a powerful disclosing agent for assumptions about the human."[7] Turing disclosed a capacious vision of intelligence, drawn from the human and animal world, full of logic, love, creativity, craft, laughter. In the years to follow, many pursuing forms of machine intelligence narrowed their sights considerably. Getting mechanical devices to imitate human intelligence first succeeded in arenas when humans have allowed themselves to behave like machines, as with algorithmic rules of production or in playing simple rule-based games. Along the way, the very notion of intelligence at issue lost much of the capaciousness Turing suggested. And in parallel, data and experience lost their centrality for the creation of intelligent behavior. How—and why?

The new computers emerging in the wake of World War II combined numerical calculation, information processing, and the manipulation of symbols according to logical rules.

Atomic bomb makers celebrated calculation; industry and cryptographers celebrated data processing; and still others focused rather on logic. A key faction in postwar machine intelligence insisted that human intelligence was most characterized by logical, symbolic thinking, and not in the lower capacities of working from sense experience (data) or in performing abundant calculations. In the mid-1950s, the most avid partisan of the logical side—a young mathematician named John McCarthy—was concerned that researchers focused on data were holding too much sway. He explained that using data would not create intelligent behavior: "the direct application of trial-and-error methods to the relation between sensory data and motor activity will not lead to any very complicated behavior." More complicated behavior would emerge only by abstracting away from sensory data.[8] Something needed to be done to get machine intelligence back on the right path.

Wait a second, we hear you cry. Why would a scientist be against data? If one side of wartime scientific activity led to a secret data-intensive state, another led to the flourishing of ideas that computers might emulate human intelligence understood more narrowly as symbolic reasoning couched in rules programmed into computers, not as inferences from data.

Against Data: Mathematics without Measurement

Applied statistics based on large amounts of data was central to the fight in World War II. Paradoxically, in the wake of the war, the hearts and minds of many scientists were won by visions of making the social sciences into subjects more like abstract pure mathematics rather than by visions of understanding society using data. Science historian Alma

Steingart explains, "It was not measurement and quantification that characterized the mathematization of the social sciences after the war, but axiomatization."[9] Reflecting on the cutting edge of social sciences, the great French anthropologist Claude Lévi-Strauss, for example, argued in 1954 that those studying human beings needed to escape quantification; they needed to set aside the amassing of data in favor of an abstract mathematical and logical treatment. Of the new mathematics of human beings, he wrote that "[t]he field with which it is concerned is not that of the infinitesimal variations revealed by the accumulations of vast accumulations of data." In fact, the study of human beings ought to be "resolutely determined to break away from the hopelessness of the 'great numbers'—the raft to which the social sciences, lost in an ocean of figures, have been helplessly clinging."[10] Lévi-Strauss complained that the social sciences have "simply borrowed quantitative methods which . . . are regarded as traditional and largely outmoded." The new "qualitative mathematics" shows "a rigorous treatment no longer necessarily means recourse to measurement."[11]

In an earlier chapter we saw how statistics moved away from its focus on the collection of data to the creation of mathematical models; so too, in the wake of World War II, did enterprises seeking to be sciences, such as sociology, economics, and political science, turn away from a dominant focus on generalizing from empirical data toward the seeking of more general, simplifying, abstract theories. In the wake of World War II, mathematical and logical theories—of human decision-making, of the economy, of intelligence—were prized and celebrated. Accumulating data, for all its importance, paled next to generalized theories, particularly those presented in abstract mathematical terms.

Thinking was far too important to be reduced to quantification. Across a swath of fields, researchers advocated the idea that humans were rational hypothesis formers, programmed by policies, not driven by data. These debates also concerned what most characterized science. And they concerned what was most distinctive about human beings. These different views offered radically different ideas of what real science *is*. And of what humans *are*.

Rules-based or symbolic artificial intelligence swam in just these anti-statistical seas. Understanding language or thinking did not require the accumulation of vast data—in fact such data would probably get in the way. Understanding—and emulating—human intelligence required abstraction and "schemas." It needed axioms and rules. It didn't need data-driven algorithms.

As we will see, computational statistics and data did not disappear. But inference from data decidedly wasn't the goal of what came to be called AI in its first decades. This anti-statistical bent characterized AI for almost half a century. A definition of AI from 1984 explained that the field deals "with symbolic, nonalgorithmic methods of problem solving," as "most of person's knowledge of a subject like medicine is not mathematical or quantitative." Rather than "mathematical or data-processing procedures," the methods involve "qualitative reasoning techniques," and "theoretical laws and definitions."[12] In other words, rules, not data.

Confecting "Artificial Intelligence"

A passionate advocate of symbolic approaches, the mathematician John McCarthy is often credited with inventing the term "artificial intelligence," including by himself: "I invented the term artificial intelligence," he explained, "when we were trying to get money for a summer study" to aim at "the long term goal of achieving human level intelligence." The "summer study" in question was titled "The Dartmouth Summer Research Project on Artificial Intelligence," and the funding requested was from the Rockefeller Foundation. At the time a junior professor of mathematics at Dartmouth, McCarthy was aided in his pitch to Rockefeller by his former mentor Claude Shannon. As McCarthy describes the term's positioning, "Shannon thought that artificial intelligence was too flashy a term and might attract unfavorable notice." However, McCarthy wanted to avoid overlap with the existing field of "automata studies" (including "nerve nets" and Turing machines) and took a stand to declare a new field. "So I decided not to fly any false flags anymore."[13] The ambition was enormous; the 1955 proposal claimed "every aspect of learning or any other feature of intelligence can in principle be so precisely described that a machine can be made to simulate it."[14] McCarthy ended up with more brain modelers than axiomatic mathematicians of the sort he wanted at the 1956 meeting, which came to be known as the Dartmouth Workshop.[15] The event saw the coming together of diverse, often contradictory efforts to make digital computers perform tasks considered intelligent, yet as historian of artificial intelligence Jonnie Penn argues, the absence of psychological expertise at the workshop meant that the account of intelligence was "informed primarily by a set of specialists working outside the human sciences."[16] Each participant saw the roots of their enterprise differently. McCarthy reminisced, "anybody who was there was pretty stubborn about pursuing the ideas that he had before he came, nor was there, as far as I could see, any real exchange of ideas."[17]

Like Turing's 1950 paper, the 1955 proposal for a summer workshop in artificial intelligence seems in retrospect incredibly prescient. The seven problems that McCarthy, Shannon, and their collaborators proposed to study became major pillars of computer science and the field of artificial intelligence:

1. "Automatic Computers" (programming languages)
2. "How Can a Computer be Programmed to Use a Language" (natural language processing)
3. "Neuron Nets" (neural nets and deep learning)
4. "Theory of the Size of a Calculation" (computational complexity)
5. "Self-improvement" (machine learning)
6. "Abstractions" (feature engineering)
7. "Randomness and Creativity" (Monte Carlo methods including stochastic learning).

The term "artificial intelligence," in 1955, was an aspiration rather than a commitment to one method. AI, in this broad sense, involved both discovering what comprises human intelligence by attempting to create machine intelligence as well as a less philosophically fraught effort simply to get computers to perform difficult activities a human might attempt.

Only a few of these aspirations fueled the efforts that, in current usage, became synonymous with artificial intelligence: the idea that machines can learn from data. Among computer scientists, learning from data would be deemphasized for generations.

Most of the first half century of artificial intelligence focused on combining logic with knowledge hard-coded into machines. Data collected from everyday activities was hardly the focus; it paled in prestige next to logic. In the last five years or so, artificial intelligence and machine learning have begun to be used synonymously; it's a powerful thought-exercise to remember that it didn't have to be this way. For the first several decades in the life of artificial intelligence, learning from data seemed to be the wrong approach, a non-scientific approach, used by those who weren't willing "to just program" the knowledge into the computer. Before data reigned, rules did.

For all their enthusiasm, most participants at the Dartmouth workshop brought few concrete results with them. One group was different. A team from the RAND Corporation, led by Herbert Simon, had brought the goods, in the form of an automated theorem prover. This algorithm could produce proofs of basic arithmetical and logical theorems. But math was just a test case for them. As historian Hunter Heyck has stressed, that group started less from computing or mathematics than from the study of how to understand large bureaucratic organizations and the psychology of the people solving problems within them.[18] For Simon and Newell, human brains and computers were problem solvers of the same genus.

> Our position is that the appropriate way to describe a piece of problem-solving behavior is in terms of a program: a specification of what the organism will do under varying environmental circumstances in terms of certain elementary information processes it is capable of performing. . . . Digital computers come into the picture only because they can, by appropriate programming, be induced to execute the same sequences of information processes that humans execute when they are solving problems. Hence, as we shall see, these programs describe both human and machine problem solving at the level of information processes.[19]

Though they provided many of the first major successes in early artificial intelligence, Simon and Newell focused on a practical investigation of the organization of humans. They were interested in human problem-solving that mixed what Jonnie Penn calls a "composite of early twentieth century British symbolic logic and the American administrative logic of a hyper-rationalized organization."[20] Before adopting the moniker of AI, they positioned their work as the study of "information processing systems" comprising humans and machines alike, that drew on the best understanding of human reasoning of the time.

Simon and his collaborators were deeply involved in debates about the nature of human beings as reasoning animals. Simon later received the Nobel Prize in Economics for his work on the limitations of human rationality. He was concerned, alongside a bevy of postwar intellectuals, with rebutting the notion that human psychology should be understood as animal-like reaction to positive and negative stimuli. Like others, he rejected a behaviorist vision of the human as driven by reflexes, almost automatically, and that learning primarily concerned the accumulation of facts acquired through such experience. Great human capacities, like speaking a natural language or doing advanced mathematics, never could emerge only from experience—they required far more. To focus only on data was to misunderstand human spontaneity and intelligence. This generation of intellectuals, central to the development of cognitive science, stressed abstraction and creativity over the analysis of data, sensory or otherwise. Historian Jamie Cohen-Cole explains, "Learning was not so much a process of acquiring facts about the world as of developing a skill or acquiring proficiency with a conceptual tool that could then be deployed creatively."[21] This emphasis on the conceptual was central to Simon and Newell's Logic Theorist program, which didn't just grind through logical processes, but deployed human-like "heuristics" to accelerate the search for the means to achieve ends. Scholars such as George Pólya investigating how mathematicians solved problems had stressed the creativity involved in using heuristics to solve math problems.[22] So mathematics wasn't drudgery—it wasn't like doing lots and lots of long division or of reducing large amounts of data. It was creative activity—and, in the eyes of its makers, a bulwark against totalitarian visions of human beings, whether from the left or the right.[23] (And so, too, was life in a bureaucratic organization—it need not be drudgery in this picture—it could be a place for creativity. Just don't tell that to its employees.)

McCarthy and Common Sense

Organizational logics were not John McCarthy's bag. Logic and common sense were. And particularly creating programs that could combine logic and common sense to achieve everyday goals. McCarthy's logical program met scathing criticism. At the 1958 conference, "Mechanisation of Thought Processes" in Teddington, London, Oliver Selfridge ridiculed the focus on deductive logic—calling "a lot of nonsense" the "notion of deductive logic being something sitting there sacred which you can borrow for particularly sacred uses and producing inviolable results." To illustrate the gulf between logic and everyday reasoning, he engaged in a shockingly misogynist invocation of the work of women. "Most women have never inferred it, but they get on perfectly well, marrying happy husbands, raising happy children, without ever using deductive logic at all." Another critic at the meeting continued with this sorry line of reasoning, to underscore how a woman learns through feedback mechanisms, not logical deductions: "If she drops the baby in a disastrous way, she does not get another chance or she gets a great yelp. She learns very quickly by crude techniques of how to achieve precise control. There is direct feedback! If she is trying to win a spouse and tries a move which does not get the right response, she quickly changes her tack."[24] The reference to women's knowledge did considerable argumentative work, resonating with later feminist critiques of AI.[25] Selfridge and others were interested in the knowledge and intelligence of everyday people, men and women alike, whereas McCarthy, an heir to the logical tradition, aimed for the ethereal knowledge of axiomatic mathematics, implicitly here the domain of men like himself and his collaborators.

What of computations using large sets of data? They didn't disappear, but they weren't really artificial intelligence as the term was then being used. After describing some statistical methods in a landmark review of the literature in 1961, Marvin Minsky argued, "I am not convinced that such 'incremental' or 'statistical' learning schemes should play a central role in our models." He admitted that such techniques "will certainly continue to appear as components of our programs" but really only by "default." True intelligence lay elsewhere: "The more intelligent one is, the more often he should be able to learn from an experience something rather definite; e.g., to reject or accept a hypothesis, or to change a goal."[26]

For all this, McCarthy and other like-minded scientists privileged mathematical and managerial modes of reasoning and acting over the much broader range of potential human knowledge to be emulated. Their approach focused on "programmed instructions operating on formal symbolic representations. . . . From the mid-1950s to the mid-1980s, it was the dominant (though not the only) approach in AI."[27] This vision for AI rested on a hierarchy of knowledge; lots of things could potentially be considered part of intelligence. These foundational figures in AI dramatically narrowed what parts of human activity could plausibly be emulated by machines as well as which parts they thought tractable for doing so. The historian Jon Agar has argued, "computerization" in the middle of the twentieth century "only took place where there were existing material practices of computation" to build upon—already existing ways of counting and classifying and organizing business.[28]

Making machines standardized enough to program to perform logical tasks was no mean feat: central to computing in the 1950s was the creation of programming languages, compilers and tools enabling people to write programs that did not depend on the idiosyncrasies of particular machines. The work, most famously embodied by Grace Hopper's creation of the first compiler, made plausible machines that active scientists could program and that could perform logical operations and data processing.[29] As computer historian Stephanie Dick notes, using the example of Simon and Newell, in actually implementing their problem solver "the programmers had to accommodate the affordances of the computer and, in so doing, abandon to an extent their commitment to simulating human practice."[30] In the chapters that follow, we will see how the challenges of implementing in actual computers with real limitations are central to the development of and distinctiveness of the data sciences.

Funding the AI Establishment

Funding long bedeviled AI before the data revolution, when the deep pockets of technology and venture capital firms opened and greatly complemented military and civilian government funds. From the start, private and public funders raised doubts. When McCarthy first approached the Rockefeller Foundation, its officers were unenthusiastic until the far more established Shannon joined, and then only gave half of the monies requested. In the United States, most of the funding during the 1970s came from various facets of the Department of Defense. Simon and Newell, working with and at RAND, drew heavily upon funds from the Air Force and Office of Naval Research. DARPA, the Defense Advanced Research Projects Agency, funded McCarthy for decades, along with a variety of other researchers in close constellation with the Dartmouth figures (and DARPA continues to fund advanced AI to this day, playing an instrumental role for years in the development of self-driving cars, as one highly visible example). In the US, AI was through and through a product of the national security state, part of a diffuse strategy of investment in technologies of potential military and commercial use—though sometimes very distant from any use. With this funding came the creation of a small community of researchers at universities and defense institutes that coalesced largely around a symbolic AI with a narrow purview of what forms of intelligence were worth pursuing.

This funding waxed and waned, in the face of great over-promises and bitter criticism. In 1969, the Mansfield Amendment required that military funding have more proximate military potential than before, putting into question much of the government's largess. In 1973, the British applied mathematician James Lighthill issued a sharply critical report on the state of AI research. Describing the success of symbolic AI, Lighthill noted, with barely disguised condescension, "problem solving in these abstract play situations has produced many ingenious and interesting programs." These successes rested upon integrating "a really substantial quantity of human knowledge about the particular problem domain." And for all the interest to psychologists, "the performance of these programs on actual problems has always been disappointing."[31] While the significance of the report is often overstated, it captured the decline in enthusiasm for highly general forms of artificial intelligence problem-solving.

The BBC broadcast a televised debate in wake of the

Lighthill report, in which McCarthy and Michie, Turing's collaborator at Bletchley, took part as defenders of Turing's dream and the nascent field. Funding in the United Kingdom waned, and resentment among less well-funded researchers in the States increased, as they voiced frustration with the failed promises of the AI founders.

In their stead came attempts to replicate more specialized forms of human intelligence.

Expert Systems

By the mid-1970s, artificial intelligence research had undergone a shift from attempting to replicate human intelligence in a general way to attempting to replicate expert knowledge.[32] Not only the code had changed. The very idea of *who* had intelligence and *what* that intelligence looked like had shifted, away from generalized capacities to narrow but deep expertise. Rather than attempting to replicate genius generalists, replicate specialized experts. The focus remained on creating rules; however, instead of general rules of intelligence, specific rules of great experts. Three major Stanford researchers, for example, concluded that the behavior of human problem solvers is "weak and shallow, except in the areas in which the human problem-solver is a specialist."[33] In 1971 Marvin Minsky and Seymour Papert argued, "a very intelligent person might be that way because of specific local features of his knowledge-organizing knowledge rather than because of global qualities of his 'thinking' which, except for the effects of his self-applied knowledge, might be little different from a child's."[34]

With this dramatic rethinking about humans came a dramatic rethinking about what to attempt using machines: "The fundamental problem of understanding intelligence is not the identification of a few powerful techniques, but rather the question of how to represent *large amounts of knowledge in a fashion that permits their effective use and interaction.*[35] The challenge then was how to move specialized expertise into a computer, in the creation of "expert systems."

Notable successes included attempts to formalize the judgment of scientists concerning organic chemical structures, as in the case of the expert system DENDRAL, created by a collaboration between the computer scientists Edward Feigenbaum, Bruce Buchanan, and the biologist Joshua Lederberg.[36] The crowning glory of this effort perhaps came with MYCIN, which automated the process of identifying bacteria in order to ensure that physicians prescribe appropriate antibiotics.[37]

The Knowledge Acquisition Bottleneck

Alas, these expert systems proved very labor intensive to create. The gulf between the contingent, complex world of medicine or industrial production and the narrow rules required by computers is vast. Experts with clinical knowledge for navigating it, it turned out, don't operate with conscious decision rules like those of computers. Figuring out the rules of experts was hard and very expensive, and the rules often proved anything but simple or concise.

So, by the early 1970s, many AI practitioners struggled to overcome the challenge of converting human expertise into "knowledge bases" and formal inference rules. Artificial intelligence researchers dubbed this fundamental difficulty the "knowledge acquisition bottleneck."[38] However good experts may be at performing actions or making judgments on the basis of sense perceptions, they all, from art connoisseurs to physicists, struggle to explain their expertise, much less to put it into the explicitly stated rules required by computers. Think only of how much background information is required to understand a simple recipe. To brown meat, for example, largely means rendering it gray through heat on a saucepan. The Australian researcher J. Ross Quinlan noted that an expert trying to explain their rules is "called upon to perform tasks that he does not ordinarily do, such as setting down a comprehensive roadmap of a subject."[39] Donald Michie, described in a 1985 interview in the journal *Expert Systems* as "one of the most prominent spokesmen for expert systems," was nonetheless "sounding a note of caution . . . recently," warning that to succeed at understanding expertise we need to have a different vision of its very nature: "Mastery is not acquired by reading books—it's acquired by trial-and-error and teacher-supplied examples. That is how humans acquire skill." Michie noted how this required a dramatically different conception of what humans are as knowers:

> People are very reluctant to accept this. Their reluctance tells us something about the philosophical self-image that we, as thinking beings, prefer. It tells us nothing about what actually happens when a teacher or a master trains somebody. That somebody has to regenerate rules from example to make them an intimate part of his intuitive skill.[40]

Early AI aspired to emulate general problem-solving; expert systems sought to emulate highly expert behavior; later expert systems were built on a different vision of the knowledge of human beings: it's often an embodied practice, very challenging to put into rules. Creating quantitative ways of predicting the skilled judgment of experts proved central to the creation of data-centric artificial intelligence. Doing that well, however, proved the death of rules. In attempting to make algorithmic the production of symbolic rules from data about expert activity, researchers by the 1990s had created forms of machine learning that, while they succeeded at predicting, failed to produce the desired concise rules. As we will see, simple rules turned out not to be the path to winning at prediction.

Even as expert systems proved to have real success in academic and commercial applications, they had become vastly specialized and lacked the resilience characteristic of human reasoners in the face of the unfamiliar.[41] The critics noticed. One argued:

> Overall, we can say that expert systems enhance their pragmatic applicability by narrowing the traditional goals of artificial intelligence research substantially, and by blurring the distinction between clever specialized programming and use of unifying principles of self-organization applicable across a wide variety of domains. This makes their significance for future development of deeper artificial intelligence technologies entirely debatable in spite of their hoped-for pragmatic utility.[42]

Much to the dismay of the expert system community, the author of these lines, Jack Schwartz, was appointed director of the Information Systems Technology Office (ISTO) within DARPA, the division which had previously (under the name IPTO) provided copious funding to the developers of AI.

Back to Bletchley Park, Back to Data

In 1959 scholars from both sides of the Iron Curtain came to the National Physical Laboratory in the UK to set out the agenda for machine intelligence and "automatic programing." The papers occasioned bitter disagreement, as when one commentator quipped that John McCarthy's paper "belongs in the Journal of Half-Baked Ideas."[43] Amid these dreams of logical and seeing machines, the English mathematician Max Newman, a former teacher and later colleague of Turing, spoke on the apparently bland subject of mechanizing what he called "more complicated clerical processes" such as determining wages and organizing library information.

Though he could say nothing of it, Newman was subtly channeling the lessons of Bletchley Park, where he had set up an operation known as the Newmanry.[44] During the war Newman had devised a technique to decrypt German codes through the large scale statistical analysis of enormous amounts of ciphertext; more than that, Newman had helped spur creation of the pathbreaking Colossus specialized computers to undertake this analysis, alongside many future luminaries of British artificial intelligence and statistics. He built up a computer operation at Manchester after the war, convincing other Bletchley Park alumni such as Turing and the statistician Jack Good to join him. Just as statisticians laundered the mathematical lessons of Bletchley Park and the NSA with biological and medical examples, Newman generalized the lessons of doing cryptography on large data sets. "It is evident," he wrote, "that a great deal of data-processing involves the recognition of pattern, and judgement as to whether patterns are alike or not."[45] Newman put finding patterns in large data sets at the heart of learning—and noted the need for vast data stores to accomplish the task.

There would seem to be no good reason why a digital computer could not be programmed to be an efficient learner machine. It would either have to be fed initially with a great amount of information about the interconnections and probabilities between the symbols of man, or else it would have to pick these interconnections up by acting as a learner machine in at least as wide a context and for as many problems, as does man. To do this it would have to have tremendous storage capacity.[46]

The academic AI community in the US and UK largely ignored these ideas of learning from data, so precious to the denizens of Bletchley like Turing and Newman. Many others working at war, manufacturing, and commerce did not.

Research into data continued "behind the fence" in the intelligence community where pattern recognition was being developed as an applied computational statistical field, funded in large part by the military to identify objects in image data. Around the same time as the Lighthill report filleted AI, the book *Pattern Classification and Scene Analysis* by the SRI International electrical engineers Richard Duda and Peter Hart introduced students and researchers to what would become fundamental ideas of machine learning, including the supervised and unsupervised learning frameworks.[47] Instead of dying, this narrower, but ultimately more powerful, form of AI thrived in industry, particularly in elements with strong military-industrial ties.

Data exploded from the 1950s, and so did efforts to understand that data. At the time, few thought of these efforts as "AI." But it was this data-driven approach to AI which has come to make our present, for better and worse, possible.

CHAPTER 8

Volume, Variety, and Velocity

The hot IT buzzword of 2012, big data has become viable as cost-effective approaches have emerged to tame the volume, velocity and variability of massive data.

—Edd Dumbill, "What Is Big Data?" 2012

I n the summer of 1953, on a flight from LA to NYC, an IBM salesman, R. Blair Smith, found himself seated next to an unkempt passenger: "his white shirt should have been changed a couple of days ago. He also needed a shave." His slovenly seatmate proved to be the president of American Airlines, C. R. Smith. The Smiths (the executives, not the band) got to chatting about the airline's struggle with managing reservations data across its network and IBM's new digital tools for data processing. "I told him," the IBM salesman explained, about "a computer that had the possibility of doing more than just keeping availability" on individual flights. It could store granular data on passengers: "It could even keep a record of the passenger's name, the passenger's itinerary, and, if you like, his phone number. Mr. C. R. Smith just was intrigued by this. And, you know, he was a true entrepreneur."[1] Working for the military and NSA, IBM was knee-deep in developing new equipment for dealing with real-time data collected by vast networks of sensors; and it was looking for the opportunity to transfer these frontier technologies into commercial applications, ideally very profitable ones.[2] The hope was to get major commercial customers to underwrite its R&D on new hardware and software development, just as the military and intelligence services had been doing. IBM sought to transfer capacities created for dealing with large real-time data about potential enemy aircraft into technologies for dealing with large real-time data about potential customers.

"Collecting data and filling seats": so read the subtitle to a description of American Airlines' SABRE System that developed in the wake of this conversation.[3] The product of ten years of development, the reservation system was an early commercial solution to the challenging problems involved in real-time processing of distributed networks of data production and decision-making. SABRE (Semi-Automatic Business Research Environment) drew upon the lessons of the government's spectacularly expensive and failed attempt to build a networked air-defense system called the SAGE (Semi-Automatic Ground Environment). Created at the intersection of academia (MIT), industry (IBM), military-sponsored think tanks (RAND), and the newly independent Air Force, SAGE involved automated record keeping, high-quality displays, and real-time networking.

In the four decades following World War II, the scale of data collected about citizens and consumers skyrocketed, as did the number of different institutions collecting it. By the end of the 1940s, the military branches charged with signals intelligence and their corporate military contractors could process streams of data computationally. A decade later, the early digital computer UNIVAC was powering the US Census, and private companies were providing the US Navy's cryptographers along with the nascent NSA the ability to "scale up" what had been previously doable only with punch cards and human labor. The US Privacy Protection Study Commission in 1977 mused that the "change in the variety and concentration of institutional relationships with individuals is that record keeping about individuals now covers almost everyone and influences everyone's life, from the business executive applying for a personal loan to the school teacher applying for a national credit card, from the riveter seeking check-guarantee privileges from the local bank to the young married couple trying to finance furniture for its first home."[4] Questions around the accumulation of data and its role in evaluation of individuals impacted far more than privacy narrowly conceived. It raised fundamental questions about access to the processes of decision-making based on that data and the means for redress—questions about who has the power to make decisions on the basis of data and who can question it.

Following the explosion of data collection around the 1970s, critics began asking key questions about the effects of data collection on privacy and justice. As we will see, many legal and political questions were set aside in the 1980s and 1990s, and many conversations around privacy were reduced to an emaciated shell, disconnected from questions of private power, and focused on fear of government rather than of private industry. The velocity of the expansion of the use of data far outpaced broad recognition of its potential harms from the mid-1990s through the 2000s. The current debate over the use of commercial data since 2010 sees the return of a conception of privacy and justice as intimately connected.

Scale, States, and Corporations

In August 1950, a muckraking column syndicated across the United States revealed that a group of Navy officers helped create a company to pursue a most secret project. And that soon, "the same Navy officers who had made the deal turned up as highly salaried vice presidents of the company." Before taking their "cushy" jobs at the company, the officers worked for the predecessors of NSA; the project was the first general electronic digital computer for cryptography, built by the new Minnesota company Engineering Research Associates (ERA).[5] Before long, the company began selling a commercial version of the machine, minus one key instruction that would disclose its cryptographic purposes. Potential impropriety aside, most developments in computing at the time came from such closely intertwined commercial and military work, characteristic of the state-driven capitalism of the Cold War.

While supporting the development of new digital computers capable of larger data storage first with ERA and then with IBM, the National Security Agency organized key early conferences to encourage industry to develop robust database solutions. The dominant company in business information processing in the mid-twentieth century, IBM, got into the digital computer business a bit late. While we would tend to see the new general-purpose electronic computers as radically different from punched card processing equipment, they initially served similar administrative tasks as older machines and dealt with similar concerns in administrative organization.

Big data needs big infrastructure—and that infrastructure had to be funded, invented, and maintained. Primarily through the military, the US government funded more than half the research and development costs of computers through the end of the 1950s, and government researchers were intimately involved in the development process.[6] Successful computers built for simulating atomic explosions and cracking codes quickly became available in commercial versions. Even horrifically expensive flops, like the SAGE defense system, channeled serious money into the development of technologies central to subsequent developments, including CRT screens and networking technologies. The NSA likely had the first transistorized computer[7] and, given its demand for data storage and processing, substantially underwrote the development costs of storage media such as automated tape systems that enabled near real-time analysis of streaming data. Once commercialized, these technologies made possible the migration of data stored on punched cards, for example, and eventually allowed for new forms of statistical analysis of that data—as well as for new forms of data to be collected.

So digital computers gained both speed in performing computations and, more importantly for our story, scale in collecting, processing, and storing data. Initially much of the work involved digitizing previously available information, but soon the capacity to capture and store information in close to real time would enable radical shifts in data collection and the running of large administrative organizations, from airlines to welfare agencies. None of this was inevitable. While many of these shifts seem obvious and predetermined in retrospect, they involved advocates and salespeople pushing organizations to partake of these new capacities, intense and always understated technological costs, and shifts in institutional logics. They involve choices about what work and knowledge matters and how they should be changed—or not. And militarism and capitalism didn't simply cause the shift to computer data processing, as these very processes transformed the military and the nature of capitalism itself by making computers central to them.

The impetus behind the technologies developed for NSA remained in the shadows for decades. But sometimes the transfer from the military to commercial applications was far more open. Early computer companies touted the potential applications for their very expensive and very high maintenance machines. A 1948 brochure for the UNIVAC asked "What's Your Problem?" and noted both data processing and computational abilities: "Is it the tedious record-keeping and the arduous figure-work of commerce and industry? Or it is the intricate mathematics of science?" The UNIVAC might be applied to "applications as diverse as air traffic control, census tabulations, market research studies, insurance records, aerodynamic design, oil prospecting, searching chemical literature and economic planning." Data collection was being normalized, and central to the pitch was lower costs in the future: "AUTOMATIC OPERATION is the key to greater economies in the handling of all sorts of information." The provision for storage allows for "extensive files and voluminous records" that can be kept "indefinitely . . . yet can be erased when no longer needed.[8] Far from hiding the government largesse and defense spending that made the machine possible, the brochure celebrates the descent of the UNIVAC from earlier functional computers, especially the "Army Ordnance Computer" (the ENIAC) and the support of the Census and Bureau of Standards.

The cold reality of converting and storing large amounts of data soon hit, a sobering reminder that data is always material and rests on dense infrastructures for its security and massive, often hidden labor for its realization. Storage tapes, for example, were so prone to fold that the problem was hidden under the technical term "dolf," solved largely through a less judicious use of lubricant and tuning of power applied to the reels.[9] Historian Janet Abbate has underscored that publicity around new computing technologies downplayed the human labor needed to make them function. Calculations about labor savings, as with a claim that the ENIAC could do twenty-five man-months of work in two hours, Abbate has argued, "included neither the 'feminine work' of preparing programs nor the 'masculine' work of machine maintenance."[10]

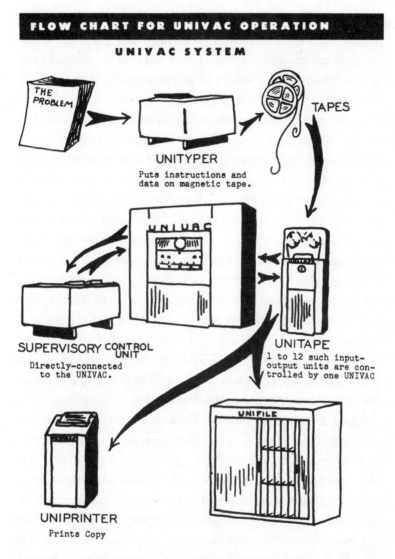

UNIVAC Advertisement. Eckert-Mauchly Computer Corporation (EMCC), 1948. Computer History Museum Archive X3115.2005. Courtesy of the Computer History Museum, Mountain View, CA.

For all the talk of "electronic brains" and computers, "electronic data processing" stuck as the term of reference for converting older systems into large-scale digital ones. While converting all sorts of business, scientific, and administrative work to digital computers may seem obvious in retrospect, each transformation involved targeted advocacy.[11] Specialists in introducing new data processing technologies to organizations emphasized the care needed to avoid resistance from employees and management alike. One early guidebook included this wisdom: "The introduction of any new system of operation faces its greatest obstacles in changing the habits of people. . . . The human problems exceed the technical problems in complexity and in difficulty."[12] Advertisements in magazines such as *Datamation* underscored the challenges organizations faced. "When forward-looking companies convert to EDP [electronic data processing] bookkeeping," a Xerox ad in 1965 noted, "transition without disruption is the big problem. Missing records, delays, and confusion can plague even the most carefully planned changeover."[13] Asking whether "the mighty data card will go the way of the abacus," an ad from Control Data Corporation in the same issue promised "new ways to keep punched cards and punch tape from coming between you and your computer."[14] And the results were far less revolutionary than expected. A 1958 report in *Business Week* noted that industry "has adopted the marvelously complex electronic computers with an almost religious fervor," and yet "often seems unsure of what to do with them." Industry is likewise "disgruntled, because early results have fallen far short of the rosy dreams in which they came wrapped." And yet the article nevertheless proclaimed the transition to these machines "inevitable," as "computers still hold the key to new systems of organization for the sprawling giants of industry, commerce and government."[15]

Corporate and government databases were indeed burgeoning by the 1960s, centralizing and standardizing data often collected on slips of papers distributed across countries, requiring arduous work. Writing about the computerization of climate models at this time, historian Paul Edwards notes, "Like all infrastructural projects, these changes involved not only scientific and technological innovation, but also institutional transformation."[16]

Credit scoring offers a key case in point: the sociologist Martha Poon has shown how credit scoring began as a bespoke activity tailored to the data individual firms collected on their customers, with highly specific models of creditworthiness. Creating credit data from transactional data was arduous and laborious, often involving women at home in a cottage industry, and long processes of punching cards.[17] Following the computerization of records, new statistical models of creditworthiness emerged, as historian Josh Lauer explains, "Computer-assisted credit scoring precipitated a fundamental shift in the concept and language of creditworthiness, even more so than computerized reporting. In addition to reducing or eliminating human contact between creditors and borrowers, scoring systems redefined creditworthiness as a function of abstract statistical risk."[18] Large-scale data collection on consumers met large-scale computation in the increasing evaluation of consumers. By the 1980s, computer modeling allowed the credit industry to produce new financial products that commodified their customers' information.[19] Computers themselves didn't make this happen: their capacities were proactively engaged and then transformed to accommodate the changing nature of these institutions.

Integrating these new technologies did not happen automatically; the pages of the trade journals and conferences from the 1960s to the 1990s teemed with pitches for redefining problems and offering solutions using data stored on new computer systems—as well as providing tips on convincing skeptical management and contending with workers and labor unions. In 1965 an LA television station aired a debate "Are Computers a Menace?" involving several luminaries from RAND discussing whether automated credit and university admissions decisions "leave the individual at the mercy of a narrow machine efficiency." The editor of *Datamation* thought not, citing "the human failure and inability to quantify all of the elements of the decision-making processes which makes it—right now—impossible to provide systems which are consistent, flexible and fair." Designers should try better. "We believe," the editor wrote, "that it *is* possible to be sane and organized about some problems which today are left to emotion, whimsy and chance. We think it's possible (and wise) to try to organize and to weigh factors which affect a decision, even if the final decision has to be left to the marvelously inefficient emotions."[20]

The Value of Information and the Revival of Privacy

"A quite different kind of electronic surveillance—and control—has become possible through the development of the giant memory machines," wrote Vance Packard in *The Naked Society* in 1964.[21] "Thus far, the information about individuals is usually fed into the super computers to serve a socially useful or economically or politically attractive purpose. But will it always be? This might especially be asked concerning those memory machines that are building up cumulative files on individual lives."[22] Packard was hardly alone. In 1976 Stanton Wheeler explained, "The very record-making process itself, then, must be regarded as problematic and we can ask not only for the conditions under which events in a person's life will become a matter of record, but whether it is legitimate for them to become a matter of record."[23] The power to record and to analyze was often necessary, but also dangerous in government and commercial hands. Those building databases needed to consider how to be certain that the destruction of privacy proved beneficial to mankind. The prophetic Packard was ahead of developments in collecting and in analyzing data.

By 1971, the economic value of data on individuals was becoming ever clearer. "The new information technologies seem to have given birth to a new social virus—'data-mania,'" Harvard law professor Arthur Miller wrote. "We must begin to realize what it means to live in a society that treats information as an economically desirable commodity and a source of power."[24]

In the wake of Watergate and revelations about illegal domestic intelligence activities, Republican senator Barry Goldwater and Democratic senator Sam Ervin aimed to establish control of personal data as a right of every American citizen. Their bill sought to restrict the violations of privacy from federal and state governments as well as private corporations. It proposed to secure the following rights to US persons:

1. There must be no personal data system whose very existence is secret.
2. There must be a way for an individual to find out that information about him is in a record and how that information is to be used.
3. There must be a way for an individual to correct information about him, if it is erroneous.
4. There must be a record of every significant access to any personal data in the system, including the identity of all persons and organizations to whom access has been given.

5. There must be a way for an individual to prevent information about him collected for one purpose from being used for other purposes, without his consent.[25]

In this framework, database builders had to be accountable to individuals. Citizens ought to know what data is being collected and who is using it for what purposes. And they should be able to put the kibosh on the collection and movement of data. Their ambitious bill eventually was narrowed to cover only the collection and use of data by federal government agencies.

In the 1960s, plans for a national centralized federal database had stoked tremendous privacy fears that led to the abandonment of the project.[26] Only a few years later a more insidious threat was apparent. Goldwater and his allies saw the dangers posed by the explosion of countless smaller databases: "[We] are building today the bits and pieces of separate automated information systems in the private and government sectors that closely follow the pattern to the present integrated communication structure."[27]

The algorithmic techniques that fuel platforms like Netflix or Facebook today were in their infancy in the 1970s, but the potency and dangers of algorithmic techniques of statistical inference from personal data were already becoming clear. Congressman Victor Veysey explained in 1974 the balance between legitimate use and personal control needed. "There is a need to develop statistical data to interpret the socioeconomic trends that continually mold the culture of this Nation, but there is a fine distinction to be drawn between data collected for justifiable purposes and the secondary purposes for which the data is sometimes used." He continued, "We must not severely restrict the legitimate services performed by" credit scoring and life insurance; "yet, we must develop adequate controls whereby information on an individual[']s personal affairs cannot be bought and sold indiscriminately."[28] The largely free exchange of data so familiar today was no natural state of affairs, and our science, laws, and regulations ought to take that into account.

Moreover, civil libertarian legislators in the early 1970s recognized that corporate and government invasions of privacy often circled around questions of race, sexual preference, and putative moral character. Invasions of privacy did not affect all people equally: the accumulation of data enabled consequential and often discriminatory gatekeeping.

In response to questioning from Senator Sam Nunn, the scholar Alan Westin was explicit: not asking for certain personal information would come at financial cost.

> if it costs us $2 more in a premium a year so that the companies will not be able to claim on an actuarial basis that they have to exclude persons who live out of wedlock or homosexuals from their rate base, I think that is a bearable cost to the American public and one, if put to them, they would probably accept. That is, Senator, by paying $2 more I won't have people asking about my sex life and checking with my neighbors and doing reports with my fellow employees as to what my sex life is. I think a lot of Americans would be willing to pay $2 a year more and not have that aspect of their lives investigated and reported on.[29]

Then as now, the free circulation of information lowered some financial costs. But at a very great cost indeed to personal lives. A key nongovernmental study put it clearly: "Privacy must be weighed today against the value gained from the collection and availability of information at central points or data banks. . . . How much personal information is worth the convenience of a credit card?"[30]

Business had a clear answer. In the wake of the introduction of the expansive bill, the complaints of industry came fast and furious, from banks, direct marketers, magazine publishers. In near unison they insisted that balancing the right to privacy with the "freedom of information" needed for business meant prioritizing the latter. Particularly galling were requirements that people consent to new uses of their data:

> We object to legislation prohibiting the transfer of information concerning individuals without the prior informed consent to those individuals. . . . Modern technology permits credit grantors to respond to consumers efficiently and rapidly partially by virtue of accessing credit information through on-line terminal facilities or alternatively by telephone inquiries. If the free flow of information is impeded by law, the resulting inefficiencies will necessarily be translated into higher costs to industry and consumer.[31]

And corporations and their fellow-traveler think tanks complained that keeping detailed records about the use and transfer of personal information was impractical and onerous. Like many other corporations, the storied retailer Sears complained, "Also extremely costly for Sears would be the requirement of maintaining a complete and accurate record of every access and use made of any data in a system, including the identity of all persons and organizations to which access has been given."[32]

In the course of the 1970s debates over the privacy bills, chamber of commerce politicians and lobbyists overcame those focused on civil liberties, and the ambitious proposed Goldwater-Ervin bill narrowed to focus exclusively on federal government collection and use of information. The enacted Privacy Act of 1974 sought to redress the balance between the interest of individuals to control information and the interest of the federal government to control and use that information. The enacted bill excised the provisions for the collection, distribution of or use of data in the private sector and substituted a commission to investigate further for the strict regulation initially envisioned. In other words, the federal government affirmed *no general principle* of the protection of personal data, and it provided no generalized form of *accounting* for the collection, exchange, and sale of that data. Instead, following the earlier protection of credit information, Americans gained crucial but narrower protections, only within specific domains of data, most notably that of students (FERPA, passed into law 1974) and, a full two decades later, medical patients (HIPAA, passed into law 1996). The momentum of reform to provide a generalized privacy law, born of Vietnam-era suspicion of government, concerns about credit agencies, Watergate, and revelations about US intelligence agencies, was squandered.[33]

Following this failure to protect nongovernmental data, the free use and abuse of personal data *came to seem* a natural state of affairs—not something contingent, not something subject to change, not something subject to our political process and choices. This norm of mostly unrestricted data collection and use created *essential conditions* for platforms in the 2000s capitalizing on granular data about people and for governments using business data for mass surveillance.

The 1977 Privacy Protection Study Commission recognized the failure of Congress to address the danger from the corporate sector as well as the federal and state bureaucracies. The enacting of the Privacy Act (1974) meant no centralized government database in the US. Those concerns had a perverse side effect: rather than one big database to rule us all, the government produced hundreds of databases hard to survey, much less to regulate and police, each subject to different regulation. Scale mattered, for it dramatically changed the privacy implications of otherwise innocuous databases. The growth of networking intensified these dangers, as the movement of records became increasingly frictionless, the speed of linking records increased, and techniques for analyzing groups of records and individual records expanded.

Despite the accelerating pace of data collection and analysis, little changed in the years following. In testimony before the US Congress in 1984, the privacy advocate Robert E. Smith presented a diagram illustrating how private databases from educational, retail, medical, and credit rating sectors intertwined with a dizzying array of state and federal databases.[34]

Combining commercial and government data could easily reveal startling parts of an individual's personal life. Smith explained the potency of combining commercial household data and IRS data:

> *MR. GLICKMAN.* . . . Is the IRS now renting computerized lists that provide demographic profiles of various households so they can find out if I go to the movies, or the Lion D'or for dinner, or Las Vegas for a weekend, and then determine if I am not paying enough in taxes?
> *MR. SMITH.* Well, not quite that data, but they could indicate that you had a Cadillac and a Ford.
> *MR. GLICKMAN.* But could they look into, let's say, my American Express account?
> *MR. SMITH.* Not into it; the fact that you had such an account might be reflected, and the general balance that you keep, that might be in there, yes.[35]

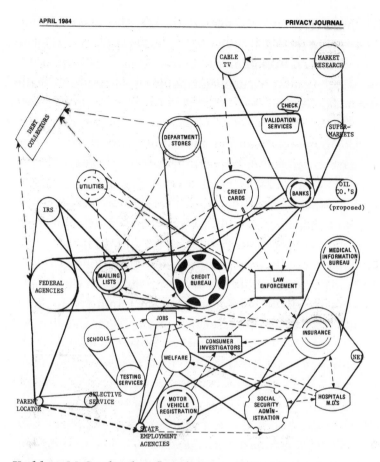

Kathleen McCarthy, data flow diagram, *Privacy Journal*,
April 1984. Robert Ellis Smith Papers, Robert S. Cox Special
Collections and University Archives Research Center, UMass
Amherst Libraries.

The pulleys in the diagram above showed Congress how
hundreds of databases had come together by the mid-1980s
to create a "de facto" national database involving most residents of the United States.

These combinations had empowered the government.

By the mid-1980s, the Congressional Office of Technology Assessment reported, "Technology has now altered that balance" between collection of data and privacy "in favor of the agencies." Combining data from multiple databases radically affected privacy.

> Computers and telecommunication capabilities have expanded the opportunities for Federal agencies to use and manipulate personal information. For example, there has been a substantial increase in the matching of information stored in different databases as a way of detecting fraud, waste, and abuse, . . . Likewise, computers are increasingly being used to certify the accuracy and completeness of individual information before an individual receives a benefit, service, or employment. . . . These technological capabilities appear to have outpaced the ability of individuals to protect their interests.[36]

How to redress the balance given the power of computers to "match" records from different databases was a pressing concern. An early example from 1977, called "Project Match," involved looking for people cheating the welfare system. "A central policy issue is whether and under what conditions the use of computer matching is appropriate, given the rights of individuals who are the subjects of matching and given the possible long-term societal effects of general electronic searches." Certain classes of people, to no surprise, find themselves subject to such matching more often: "Computer matches are inherently mass or class investigations, as they are conducted on a category of people rather than on specific individuals. In theory, no one is free from these computer searches; in practice, welfare recipients and Federal employees are most often the targets."[37]

Database technologies increasingly could do far more than matching people across multiple databases. The Privacy Protection Study Commission warned in 1977: "The real danger is the gradual erosion of individual liberties through the automation, integration, and interconnection of *many small,* separate recordkeeping systems, each of which alone may seem innocuous, even benevolent, and wholly justifiable."[38] Scale mattered, and changed the privacy implications of otherwise innocuous databases. The growth of networking intensified these dangers, as the movement of records became increasingly frictionless, the speed of linking records increased, and techniques for analyzing groups of records and individual records increased.

The balance of privacy and government interest had tilted in discriminatory ways, in ways more and more familiar today, to focus ever more on just those people least able to demand accountability, least empowered to push back and demand that systems fulfill the full range of our collective expectations for data analysis in a just democratic society.[39]

The growth of these forward-looking concerns happened before access to the internet became widely available, just as the personal computer was beginning to appear in homes and workspaces across the United States and beyond. No legislation followed. As databases expanded and became ever more ubiquitous, the everyday practice of collection and exchange of data cemented the presumption that no general principle of protection governed non-federal governmental and corporate use of personal data, with some major exceptions for health, credit, and educational data. The lack of any principle of general protection for personal data came to seem ever more natural. Rather than being seen as a political choice, the absence of privacy protection came to be mistakenly understood as the nature of data and data collection. The full significance of this failure took another two decades to become widely apparent, in the 2010s. Only then did the twin dangers of corporate and government trafficking in personal data move from a concern of small groups of activists to the front pages of newspapers and news feeds alike.

In 1999, Scott McNealy, the CEO of Sun Microsystems, insisted, "You have zero privacy anyway. Get over it."* By 2010, Mark Zuckerberg, founder and CEO of Facebook, claimed that privacy was no longer a "social norm."[40] Neither statement is true. But powerful interests worked to make these beliefs seem to many to be true.

In 1973, W. Lee Burge, head of the firm that became today's consumer credit reporting giant Equifax, argued, "Through the free flow of [personal and financial] information—by having accurate, pertinent facts at their disposal—American businessmen can and do act with the kind of confidence that keeps our economy alive and thriving."[41] Such a defense of the collection and exchange of data prioritizes innovation and economic efficiency over other human values. In a defensive response to a congressional inquiry in the 2000s into services that sell data on US persons, the major data broker Acxiom linked the collection and analysis of data to main-

---

* Stephen Manes, "Private Lives? Not Ours!," *PC World*, June 2000. To clarify: the CEO at the time, Scott McNealy, was not talking about surveillance capitalism or machine learning; he was talking about the design of a chip, relative to that of the competing Intel Pentium III chip, which could possibly expose the details of computation. Nonetheless, the pithiness and provocativeness of the dismissal of consumer protection was often repeated over the next decade, as the dominant technology changed from hardware companies to information platforms. Polly Sprenger, "Sun on Privacy: 'Get Over It,' " *Wired*, January 26, 1999, https://www.wired.com/1999/01/sun-on-privacy-get-over-it/.

taining freedom itself: "because many web applications are made available free of charge to the public, ideas are organically expressed and exchanged on a daily basis in a manner never before seen. We have recently witnessed how various social media sites have mobilized and energized citizens . . . . For these populations, information truly is a direct conduit to Liberty."[42] The price of freedom is not vigilance but being data mined: free as in beer yet again substitutes for free as in freedom.

Defenders of the free exchange and collection of data from the 1970s to the present intone about the trade-offs we must expect should we collectively choose to protect private data more robustly. Privacy, they argue, comes at a major financial cost: more expensive services and products and barriers to innovation. And it comes, others argue, at a national security cost: decreased government ability to find and neutralize nefarious forces. These politically powerful arguments, long favored in both Democratic and Republican administrations, lower and narrow our collective expectations about the use of our personal data and its significance for privacy, autonomy, and liberty.

Drawing upon long-standing free-market narratives, industry stories for decades have stressed the *absence* of the US government from innovation to justify the narrowing of our collective expectations—even though, as we have stressed, government investment created and nurtured the computer industry. When it came to protecting personal data, in this narrative, the US has had a light touch. In a report from a few years ago, one think tank explains (to take an example):

In the data economy, this meant avoiding comprehensive data-protection rules that limit data sharing and reuse, and instead focusing on developing tailored regulations for specific sectors, thereby allowing most industries the freedom to innovate. These policies formed the core regulatory environment that allowed companies from Amazon and eBay to Google and Facebook to thrive, and provided a distinct alternative to the precautionary, innovation-limiting rules Europe adopted.[43]

In this narrative, corporate freedom to collect, buy, trade, and mine data allowed the US to become the tech powerhouse it is today. In these stories, corporate rights to information flows trumped any overdrawn right to privacy to awesome effect and with the clear moral that we should today avoid robust forms of algorithmic regulation and accountability. Missing from such free-market stories are the vast federal funds—mostly defense and intelligence dollars—that made the microelectronic industry possible and midwifed the internet itself. Equally missing from these tales are our collective legitimate expectations in domains other than economic efficiency and profit of a small number of firms.

In 1969 internet pioneer Paul Baran remarked: "we may be making as much of a mistake in expecting the computer manufacturers to straighten out the privacy problems as we have made in expecting automobile manufacturers to design adequate smog-control devices of their own accord and without prodding."[44] A year earlier he argued before an audience at MIT: "Those who deal with records that can brand and divide must modify their actions toward the best long-range interests of society, even when such modification conflicts with the best interest of individual agencies or corporations. Is this too much to ask?"[45]

Emaciated Privacy

In the 1970s, arguments around privacy focused squarely on the potential harms of automatic decision making and the likely disproportionate impact upon less empowered groups. Not only a question of liberty of atomized individuals, privacy was a concern of civil rights that pertained to particular classes of people, as dossiers collected primarily on Black students well illustrated. Critics in the 1970s saw clearly that scale altered the effects of records; many people since have tried to limit our understanding of these effects to legitimate their activities. Rather than exclusively considering privacy as an individual right, advocates at the time stressed that the concept encompassed concerns about the uneven distribution of harm among social groups and their different abilities to seek justice. Toward the end of the century broader discussions of privacy in the US narrowed considerably, as did so much of the political imagination as part of the broader movement toward a libertarian conception of the individual and their rights. The Harvard philosopher Robert Nozick exclaimed, "There is no *social entity* with a good that undergoes some sacrifice for its own good. There are only individual people, different individual people, with their own individual lives."[46] Economists and policymakers alike followed Milton Friedman ever more in his vision that society has no obligation, only individuals.[47] "In an ideal free market resting on private property, no individual can coerce any other, all cooperation is voluntary, all parties to such cooperation benefit or they need not participate. There are no values, no 'social' responsibilities in any sense other than the shared values and responsibilities of individuals. Society is a collection of individuals and of the various groups they voluntarily form."[48]

Not just defanging the regulatory state through the end

of the century, these forms of emaciated social and economic thinking made it far less easy to think clearly about privacy in the age of large data. The twin focus on coercion from government and private entities gave way to what the legal scholar Jodi Short called the "paranoid style" of thinking about government regulation. Deep worry about the dangers of state coercion overwhelmed concern over private power.[49] In this libertarian world, privacy was increasingly treated narrowly as an individual civil liberty against the overextension of government. The legal scholar Priscilla Regan noted the limits of just such an individualistic approach in 1995: "Defining a problem in terms of rights has been a potent political resource for many issues—civil rights, women's rights, rights of the disabled—but these issues involve rights to some benefit or status and are defined not in terms of an atomistic individual but an individual as a member of a group."[50] While activist organizations as well as luminous critics like Oscar Gandy pushed hard against this narrowing of critique, intellectual trends and business interests alike muted the prominence in policy and even activist circles that they deserved.[51] The loss of these critiques and the technical knowledge around these broader conceptions of privacy was far from accidental. The Congressional Office of Technology Assessment provided many of the insights from the 1970s quoted above; Newt Gingrich's House of Representatives shuttered the office in 1995. At the very moment when the commercialization of the internet took off during the Clinton presidency, a government working group, building upon the work of the 1970s just discussed, warned that the internet would make the creation of profiles of individuals easy and low-cost, without the labor and travel required previously. The diagnosis was correct. At that moment, the media scholar Matthew Crain argues, however, cures were envisioned almost entirely as a question of individual choice, of empowering individual users.[52] And so was begat our world of ubiquitous profiling and of opting out individually of cookies on your machine.[53]

Even those who aimed to bring the spirit of the 1960s to the internet worked within a dramatically individualistic understanding of privacy. In fact, political advocates celebrated the coming of the internet as quite precisely undermining differences of scale. "The fundamental thing" the internet does, Esther Dyson explained, "is to overcome the advantages of economies of scale . . . so the big guys don't rule."[54] In this vision, the internet made the individualistic reveries of a Friedman *more true* not less true by liberating individuals from atavistic social bonds. Historian Fred Turner argues,

> Even as they conjured up visions of a disembodied,
> peer-to-peer utopia, . . . writers such as Kevin Kelly,
> Esther Dyson, and John Perry Barlow deprived
> their many readers of a language with which to
> think about the complex ways in which embodi-
> ment shapes all of human life, about the natural and
> social infrastructures on which that life depends,
> and about the effects that digital technologies and
> the network mode of production might have on life
> and its essential infrastructures.[55]

In the very process of celebrating and defending the internet against government intrusion, the political vision found in magazines such as *Wired* reinforced narrow conceptions of privacy as *individual* rights. And it did so in the broad moment of distrust in governments as slow and inefficient across much of the political spectrum.[56] As a result, even many activist accounts of privacy left citizens dramatically underequipped to contend with the risks that the new aggregations of data and analysis which the internet largely made possible.[57] Despite the work of many scholars, activists, and technologists to broaden the understanding of surveillance, this narrowing of political, social, and legal imagination at just the moment that the scale of data collection and analysis exploded provided inadequate philosophical and legal accounts to understand what had happened and to imagine the political and social response to automatic decision making with large data comport with our collective aspirations.[58]

Around the time of the attacks of September 11, 2001, government agencies such as the NSA and its British counterpart GCHQ had moved beyond collecting and breaking Nazi and then Soviet codes, as they became able to collect and analyze the telephone and internet use of people worldwide, including potentially their own citizens. National security lawyers, defense intellectuals, and law enforcement in the US and the UK in the late 1990s called for transformation in the laws and definitions around wiretapping, given the vast expansion of communications via the internet and mobile telephony—but lacked the political ability to actualize them over the objections of civil libertarians. In the immediate wake of 9/11, a reactive US Congress passed the PATRIOT Act of 2001, which effected subtle changes in domestic surveillance law, among its many provisions, whose full import remained hidden for years.

Hamstrung by an emaciated view of privacy, judges and policymakers alike suffered from the failure of imagination in contending with these new analytical technologies. The courts that were supposed to regulate the NSA in the wake of 9/11 illustrated a shocking lack of imagination about how scale dramatically changes the effects of the collection and processing of data. The limitations of a highly individualistic approach to privacy appeared quickly in the wake of revelations of the NSA's program of collecting the "metadata" of phone calls after the 9/11 attacks. Metadata meant only the phone numbers, not the content of the communications. The NSA and other authorities have long claimed that metadata does not share the same constitutional protections as content, and they've often convinced legislators and courts to agree. While possessing the ability to demand detailed accounts and empowered to push back against the NSA, the courts lacked the competencies to challenge the technical claims of the agency. The NSA provides vast formal accountings of its collective activities to the secret Foreign Intelligence Surveillance Court (FISC)—a tremendous degree of transparency behind closed doors, albeit unknown outside the court until recently. For all its expertise in the law, the court has lacked sufficient technical knowledge about the aggregation of data to push back effectively against these accounts.

The key court decisions establishing that communications "metadata" have a far lower level of constitutional protection than "content" rest on a series of arguments about the lack of constitutional protection afforded to an individual in the moment of making a telephone call. Under the prevailing understanding, people making telephone calls freely give to the telephone company the phone numbers they are dialing. They have no "legitimate expectation of privacy" as to that dialed phone number, even though they expect the content of the call to remain private. This is true for each call that they make. The government therefore can acquire the metadata for each call without any question of search and seizure. The lack of any reasonable expectation of privacy extends from the individual calls to any aggregation or analysis of them. Operations upon data without constitutional protection, according to this analysis, exclusively yields facts without constitutional protection.

Rulings of the secret Foreign Intelligence Surveillance Court since the mid-2000s treat this matter of aggregation plainly. Fourth Amendment rights are personal: so "long as no individual has a reasonable expectation of privacy in meta data [sic], the large number of persons whose communications will be subjected to the . . . surveillance is irrelevant to the issue of whether a Fourth Amendment Search or seizure will occur."[59] A later ruling developed the reasoning further: "Put another way, where one individual does not have a Fourth Amendment interest, grouping together a large number of similarly-situated individuals cannot result in the Fourth Amendment interest springing into being *ex nihilo*."[60]

The Fourth Amendment interest of all the individuals does not, however, arise ex nihilo as the court says; it arises from the challenges to legitimate *individual* privacy interests that mass collections of such data make possible given current analytical tools. Princeton computer scientist Edward Felten noted in an important court filing, "Sophisticated computing tools permit the analysis of large datasets to identify embedded patterns and relationships, including personal details, habits, and behaviors. As a result, individual pieces of data that previously carried less potential to expose private information may now, in the aggregate, reveal sensitive details about our everyday lives—details that we had no intent or expectation of sharing."[61] The promise of analyzing metadata to discover patterns common to terrorists rests precisely on the assumption that such analysis can uncover latent phenomena about individuals, not just aggregates. The NSA's own historians explain the growth of the agency's ability in the 1950s "to derive useful information from the externals of message traffic, in addition to or apart from success in reaching the underlying plaintext of the message contents." This ability to work with what would later be called metadata "ranks as a defining event in cryptologic history."[62] The power of computational statistics, not just statistical inferences about collectivities, but even more fundamentally to unmask often intimate and personal aspects of specific individuals, underscores that there is a deep privacy interest limiting the use of such analytical tools. Many of our older intuitions about our ability to consent knowingly and rationally to giving up information about ourselves in the era of big data are deeply wrong. Recent court rulings about surveillance, notably United States *v.* Jones and Carpenter *v.* United States, show that the judiciary is slowly replacing some of these dated technical intuitions. The power of current and future analytical tools demands institutions with the knowledge and critical power to undertake a rethinking of consent in the age of aggregation, given our lack of epistemic and ethical intuitions about the power—and dangers—of machine learning platforms.[63] Our next two chapters consider the development of these powerful analytical tools.

From Data to Optimization as Value

Storing data, for all its challenges, proved far easier than analyzing data for insight. In an era with only minor limits on the collection of data, growing corporate and governmental data presented a major technical challenge. No one knew which tools could produce meaning—and value—from these databases. Data were being collected at an increasing rate without clear means for studying—and profiting—from them.

Funders of data analysis technologies had become impatient. They had been sold many bills of goods over decades. For example, the US Census Bureau was an early adopter of technologies for contending with data—from Hollerith punched card machines to the UNIVAC discussed above. By the 1980s, its staff had become a bit exasperated with those they were funding. "For almost three decades, staff at the Census Bureau have heard claims that machine recognition of handwriting was just around the technological corner. However, a careful review of most claims showed that the corner was still a long way off."[64] With tighter constraints on budgets, and greater skepticism toward grandiose claims about artificial intelligence and allied fields like machine translation, DARPA and allied agencies created a new approach to evaluating projects, an approach that involved single metrics to score success on data given to all competitors. This became known later as a common task framework. Developments in the 1980s suggested some reason for optimism regarding reading handwriting automatically, so the Census Bureau and the National Institute of Standards and Technology (NIST) set up a competition to see where the long-promised technologies stood, to encourage progress, and to gear up rivalry among firms and academics.

The Census Bureau did not want the competition to work with "toy" data, far removed from the complexity of real-world data. The data were available thanks to a digitized handwriting sample form, where census workers and school children copied out a sequence of numbers, letters, and words into clearly demarcated boxes.

"It was decided that a test open to organizations having strong [optical character recognition] programs would be a cost-efficient tool for meeting these goals. This would allow comparison of the results from a wide variety of systems, algorithms, features, and preprocessing."[65] Teams from across the United States and Western Europe joined the fray—mostly from corporations like Eastman Kodak, the Thinking Machines Corporation, IBM's Almaden Labs, and Daimler-Benz's AEG, with a smattering of participants from universities from Michigan to Valencia and Bologna. Many of the firms adapted their commercial technologies that attempted to read addresses or checks.

The results? "About half of the systems correctly recognized over 95% of the digits, over 90% of the upper-case let-

HANDWRITING SAMPLE FORM

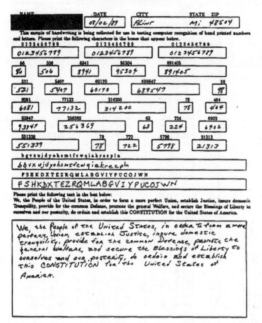

R. Allen Wilkinson, Jon Geist, Stanley Janet, Patrick J. Grother, Christopher J. C. Burges, Robert Creecy, Bob Hammond, et al., *The First Census Optical Character Recognition System Conference.* NIST IR 4912, p. 19.

ters, and over 80% of the lower-case letters in the test. For comparison, a human correctly recognized about 98.5% of the test digits."[66] NIST made available the results, as seen in the next illustration.

The systems ranged widely, from neural networks to statistical pattern recognition, from Spearman's principal component analysis to the lowly nearest neighbors algorithm (developed by statisticians in 1951, with support of the US Air Force). AT&T's Bell Labs submitted four candidate classification systems, including a variant of their commercial

Entered System	Percentage Classification Error		
	Digits	Uppers	Lowers
AEG	3.43 ± 0.23	3.74 ± 0.82	12.74 ± 0.75
ASOL	8.91 ± 0.39	11.16 ± 1.05	21.25 ± 1.36
ATT.1	3.16 ± 0.29	6.55 ± 0.66	13.78 ± 0.90
ATT.2	3.67 ± 0.23	5.63 ± 0.63	14.06 ± 0.95
ATT.3	4.84 ± 0.24	6.83 ± 0.86	16.34 ± 1.11
ATT.4	4.10 ± 0.16	5.00 ± 0.79	14.28 ± 0.98
COMCOM	4.56 ± 0.91	16.94 ± 0.99	48.00 ± 1.87
ELSAGB.1	5.07 ± 0.32		
ELSAGB.2	3.38 ± 0.20		
ELSAGB.3	3.35 ± 0.21		
ERIM.1	3.88 ± 0.20	5.18 ± 0.67	13.79 ± 0.80
ERIM.2	3.92 ± 0.24		
GMD.1	8.73 ± 0.35	14.04 ± 1.00	22.54 ± 1.22
GMD.2	15.45 ± 0.64	24.57 ± 0.91	28.61 ± 1.25
GMD.3	8.13 ± 0.39	14.22 ± 1.09	20.85 ± 1.25
GMD.4	10.16 ± 0.35	15.85 ± 0.95	22.54 ± 1.22
GTESS.1	6.59 ± 0.18	8.01 ± 0.59	17.53 ± 0.75
GTESS.2	6.75 ± 0.30	8.14 ± 0.59	18.42 ± 1.09
HUGHES.1	4.84 ± 0.38	6.46 ± 0.52	15.39 ± 1.10
HUGHES.2	4.86 ± 0.35	6.73 ± 0.64	15.59 ± 1.08
IBM	3.49 ± 0.12	6.41 ± 0.80	15.42 ± 0.95
IFAX	17.07 ± 0.34	19.60 ± 1.26	
KAMAN.1	11.46 ± 0.41	15.03 ± 0.79	31.11 ± 1.15
KAMAN.2	13.38 ± 0.49	20.74 ± 0.88	35.11 ± 1.09
KAMAN.3	13.13 ± 0.45	19.78 ± 0.60	33.55 ± 1.37
KAMAN.4	20.72 ± 0.44	27.28 ± 1.30	46.25 ± 1.23
KAMAN.5	15.13 ± 0.41	33.95 ± 1.22	42.20 ± 0.96
KODAK.1	4.74 ± 0.37	6.92 ± 0.78	14.49 ± 0.77
KODAK.2	4.08 ± 0.26		
MIME	8.57 ± 0.34	10.07 ± 0.81	
NESTOR	4.53 ± 0.20	5.90 ± 0.68	15.39 ± 0.90
NIST.1	7.74 ± 0.31	13.85 ± 0.83	18.58 ± 1.12
NIST.2	9.19 ± 0.32	23.10 ± 0.88	31.20 ± 1.16
NIST.3	9.73 ± 0.29	16.93 ± 0.90	20.29 ± 0.99
NIST.4	4.97 ± 0.30	10.37 ± 1.28	20.01 ± 1.06
NYNEX	4.32 ± 0.22	4.91 ± 0.79	14.03 ± 0.96
OCRSYS	1.56 ± 0.19	5.73 ± 0.63	13.70 ± 0.93
REI	4.01 ± 0.26	11.74 ± 0.90	
RISO	10.55 ± 0.43	14.14 ± 0.88	21.72 ± 0.98
SYMBUS	4.71 ± 0.38	7.29 ± 1.07	
THINK.1	4.89 ± 0.24		
THINK.2	3.85 ± 0.33		
UBOL	4.35 ± 0.20	6.24 ± 0.66	15.48 ± 0.81
UMICH.1		5.11 ± 0.94	15.08 ± 0.92
UPENN	9.08 ± 0.37		
VALEN.1	17.95 ± 0.59	24.18 ± 1.00	31.60 ± 1.33
VALEN.2	15.75 ± 0.32		

Table 3: Mean zero-rejection-rate error rates and standard deviations in percent calculated over 10 partitions of TD1.

R. Allen Wilkinson, Jon Geist, Stanley Janet, Patrick J. Grother, Christopher J. C. Burges, Robert Creecy, Bob Hammond, et al., *The First Census Optical Character Recognition System Conference.* NIST IR 4912, p. 9.

product based on neural networks. The researchers contributing to the Bell Labs submission included many of the future luminaries of machine learning, including Isabelle Guyon and Yann LeCun.

Optimizing a process to classify some data correctly

was a far cry from the grandiose dreams of artificial intelligence discussed in the previous chapter. The *values* the Census Bureau and NIST insisted upon involved accuracy of prediction and efficiency, not intelligibility or a grounding in a symbolic logical process; the Census Bureau and NIST were likewise concerned with speed in dealing with its real-world data at scale. Such a dramatic transformation of values is central to the subsequent explosion of machine learning and artificial intelligence. The valuing the optimization of metrics for real world applications drove much subsequent development of machine learning, data mining, and the data sciences from the late 1980s to the present. Problems like the accurate recognition of handwriting exemplify a sharpening of focus on problems characterized by a clear numerical metric of success to optimize. And character recognition likewise exemplifies an insistence of the creation of robust algorithmic systems capable of dealing with real world data, not artificially clean data, at ever-larger scales, often in real time. Changing the goals from understanding or artificially creating "intelligence" to one of maximizing quantitative performance also facilitated a competitive, community-organizing task. Such competitions are useful for organizing a community around an engineering goal of performance, however much it shifts focus away from the loftier goals of, for example, the organizers of the Dartmouth Workshop in 1956. Our next chapter sketches the controversial blossoming of pattern recognition and machine learning focused ever more squarely on values of optimizing for clear metrics of success, rather than broad, but vague, aspirations for intelligence. And the subsequent chapter, on data science, looks at the transformation of these algorithms to work at the industrial scale required to deal with the real-world data collected by corporations, scientists, and governments. And this narrowing of values to optimization? It's at the heart of the ethical and political dilemmas around AI today.

What began as the challenge of computing with data grew into a profitable industrialized data mania, far distant from the concerns of McCarthy and the other original framers of artificial intelligence—indeed nearly anathema to them. Yet, as we will see, the data mania in some ways came back to give a second or third life to artificial intelligence— AI focused squarely on learning from data, not handcrafting symbolic rules.

CHAPTER 9

Machines, Learning

P at Langley was disappointed. By 2011, the academic field he'd spent much of his life nurturing, machine learning, had exploded in influence, funding, and size. But success came at a huge cost: the field had largely given up on "more complex tasks like reasoning, problem solving, and language understanding" in favor of simpler tasks like prediction. Instead of "sophisticated systems that carried out multi-step reasoning, heuristic problem solving, language understanding, or other complex cognitive activities," the field had limited itself to simpler statistical tools designed to solve easier problems. Machine learning had moved from grand, prestigious problems of emulating human knowledge to focus narrowly on numerical prediction and classification.[1] Machine learning seemed far more ambitious when he described the same field more than a quarter century earlier, in 1984, separating the narrow goals of "pattern recognition" from the "symbolic" approach of AI: "Historically, researchers have taken two approaches to machine learning. Numerical methods such as discriminant analysis have proven quite useful in perceptual domains, and have become associated with the paradigm known as Pattern Recognition. In contrast Artificial Intelligence researchers have concentrated on symbolic learning methods."[2] In the intervening years, the values of machine learning had changed, as had its criteria for success. Ironically, just this dramatic narrowing has enabled its extraordinary success today. Critics in the 1970s and 1980s doubted artificial intelligence would amount to much of anything, whatever dystopian novelists might envision. Critics in the 2020s worried that artificial intelligence would take over nearly all domains of human decision-making, just as the dystopians warned. Through a remarkable rebranding that the most effective marketers could only dream of, the term "artificial intelligence" today has become nearly synonymous with a narrower slice of statistical techniques for making predictions called deep learning. This chapter sketches that story.

Some fields, like biology, are named after the object of study; others like calculus are named after a methodology. Artificial intelligence and machine learning, however, are named after an aspiration: the fields are defined by the goal, not the method used to get there. From the 1960s to the 2000s, researchers in machine learning took methodologies from wherever necessary (despite the scorn of many anointed scientific leaders): neural nets, the methods of "pattern recognition" from electrical engineering, and even mathematical statistics. These disparate methodological borrowings would go on to forge the AI renaissance of the 2010s to the present.

Symbolic AI Kills the Neural Net Star

In 1980, few would have expected predictive models to take over AI. Working within a highly competitive funding landscape, the devotees of symbolic AI derided more data-driven and statistical approaches. And they especially denigrated efforts to take the neural networks of a human brain as a model for machines capable of learning from perceptions. The most famous example, the Perceptron, sought to learn to discriminate among objects "seen" by artificial neurons.

Envisioned by Frank Rosenblatt in the 1950s, the Perceptron involved the effort to recognize sensory input without the hard coding of rules. Rosenblatt sought "a machine which would be capable of conceptualizing inputs impinging directly from the physical environment of light, sound, temperature, etc.—the 'phenomenal world' of light, sound, temperature—rather than requiring the intervention of a human agent to digest and code the necessary information."[3] With substantial military funding, he constructed a brain analogue—an artificial network—to recognize objects without resort to long logical processes.[4] First realized in specialized hardware, the Perceptron then became a more standardized algorithm able to run on general purpose digital computers. Rosenblatt had a flair for publicity, which may have spurred his critics. In 1958, *The New York Times* ran a story with the headline "New Navy Device Learned by Doing," and the claims were stunning, describing "the embryo of an electronic computer that it expects will be able to walk, talk, see, write, reproduce itself and be conscious of its existence."[5]

The program was a potent alternative to symbolic artificial intelligence—an alternate way of appreciating and emulating human intelligence—and critics went after Rosenblatt's program with great vehemence. By the late 1960s, however, artificial neural networks were widely perceived as a dead end. Simple neural networks can only classify objects using a linear boundary. What does this mean and why does it matter? A Perceptron is not capable of "learning" some easy logical functions of its inputs, such as so-called "exclusive or." "Exclusive or" is the "or" familiar from wedding invitations: you can choose the beef or the chicken or the tofu surprise, but not all three or even two, unless you steal from your neighbor. If being able to do symbolic logic is the mark of an intelligent system, then being unable to deal with exclusive or is a death knell.

But this limitation proved not quite as a killing strike as it seems. Researchers soon realized that adding additional layers of "neurons" beyond the first layer could indeed create nonlinear forms of classification. Neural nets could thus learn things like "exclusive or." The rub? In the 1960s and early 1970s, no one knew what algorithm could be used on lots of data to "train" a multilayer neural network in an efficient way or with any degree of confidence that the network would improve in a systematic way. In 1983, the economist and AI pioneer Herbert Simon, who had demonstrated his Logic Theorist at the Dartmouth Workshop of 1956, confidently claimed,

> the whole line of Perceptron research and nerve net learning . . . didn't get anywhere . . . those systems . . . never learned anything that people didn't already know. So they should again strengthen our skepticism that the problems of AI are to be solved solely by building learning systems.[6]

Neural networks seemed dead, way dead, except among a small number of devotees in Japan and a few other places. And many in the AI community were happy not to have competition for US military patronage and academic positions. Besides the competition for dollars, neural nets were distasteful, for they involved a radically different vision of what comprises intelligence. For critics like Simon, the evident failings of neural networks cast doubt more broadly on any attempts to build learning systems focused primarily on learning from data.

Away from the hallowed halls of artificial intelligence, however, systems trained on data didn't disappear or lose all their funding. They are rebranded at the heart of the AI of today.

Pattern Recognition, for Example

In the early 1960s, engineers at Philco, newly a division of Ford Motor Company, worked under contract with the US Army on technological means to aid the military in the automated recognition of features in photos taken by spy planes, like the U-2. Among the bevy of technologies supported was the use of computational statistics to aid classification of objects in photos. It was in just such commercial and academic labs, funded by the US military and intelligence agencies, that uses of computational statistics focused more on predictions based on data flourished. Researchers such as the Philco engineers working within the broad rubric of "pattern recognition" sought techniques to discriminate among objects, estimating parameters for known distributions, and, even more challenging, to begin the tough task of discerning probability distributions when their underlying form cannot be assumed.[7] They worked at government labs, at corporate labs, and at great universities like Cornell, USC, and Stanford, typically with copious military support.[8] No corporate lab shone as brightly as Bell Labs in New Jersey.

When they surveyed the field in the 1960s and early 1970s, researchers explained that pattern recognition involved less an academic discipline than a cluster of like-minded practitioners oriented around common sets of goals. The neural network idea of the Perceptron is perhaps the best known of these efforts. Most researchers in pattern recognition ultimately cared—and care—little whether neural networks in any way replicated human cognition: the networks were tools for prediction, not means for understanding the brain. By the 1960s, practitioners argued, pattern recognition succeeded in large part because it had abandoned the effort to simulate human perception: "Whatever successes we have had . . . have been the result of an effective transformation of a perception-recognition problem into a classification problem."[9] And pattern recognition researchers cared little about the symbolic side of artificial intelligence.

In these labs, an attitude focused on practical results from large accumulations of data flourished. In the course of this work, early forms of the key algorithms now central to the contemporary data sciences emerged and were modified to work within the computational limits of their times. This meant less theorizing about symbols or schema than devising means for implementing algorithms within limited hardware using real data sets. While these algorithms appeared in academic papers, they were primarily implemented in experimental and commercial systems. Making predictive systems with real-world data required sometimes ugly engineering. "Practical considerations of computer economics often prevent the wholesale application of the methods mentioned above to real-life situations." Such situations require "somewhat undignified and haphazard manipulation . . . to render the problem amenable to orderly solution," including "preprocessing, filtering or prefiltering, feature or measurement extraction, or dimensionality reduction."[10] Techniques for handling real-world data were integral, not ancillary, to pattern recognition in practice: no matter how elegant the algorithm, if it could not deal with large-scale data from "real-life" situations on limited disk drives and computers, then it needed to be set aside or modified.

Machine Learning Learns, from Pattern Recognition—and More

Pattern recognition was but one of the many sources of successful methods for machine learning by the end of the twentieth century. Machine learning itself was more a specification of an aspiration than of a method. The "real-world" attitudes described above came at a cost. Practitioners over the course of the late 1980s and 1990s abandoned the AI goals of simulating how human beings reason on computers or using computers to attempt to understand human cognition. Seeking out "what works" rather than what is true or beautiful encouraged a magpie-like search for algorithms and practices for making some sense of data. The field of machine learning slowly but decisively adopted these values, more of a practical engineering tradition than of the pure sciences, more aligned with industry than the academy. And machine learning researchers did so with uneven but increasing access to computational time enabling such an approach, at least in well-funded labs.[11] Animated by eclecticism, machine learning drew widely from algorithms across many fields of practice and inquiry: pattern recognition, signal processing, clustering, as well as from computationally focused statistics. Indeed, statisticians are wont to complain that machine learning keeps reinventing the wheel. Harkening back to the tradition of practical wartime statistics, most machine learning since the late 1980s involves minimizing some specified error or "loss function," which Abraham Wald put at the heart of his sequential decision theory and made its way into pattern recognition.[12] Many machine learners came to embrace Bayesian statistics, long disavowed by mathematical statisticians within the academy, though celebrated within the corridors of the intelligence community.

In a moment of profound irony, machine learning, a little-respected relative of artificial intelligence, would come in the new millennium to become the greatest success, even savior of AI, to such an extent that after 2013 machine learning came largely to displace the far more ambitious goals of traditional AI, and the terms came to be used interchangeably.

From Artificial Intelligence to Machine Learning

John McCarthy's 1955 funding proposal celebrated "the conjecture that every aspect of learning or any other feature of intelligence can in principle be so precisely described that a machine can be made to simulate it."[13] The problems in realizing artificial intelligence proved far more intractable than tasks such as targeting radar or identifying tanks. "Workers entered the field [of artificial intelligence] around 1950, and even around 1960," the highly critical report by Sir James Lighthill from 1973 explained, "with high hopes that are very far from having been realized in 1972. In no part of the field have the discoveries made so far produced the major impact that was then promised."[14] AI research had included both attempting to create intelligent behavior and better understanding human intelligence. By 1987, a commentator noted, "No one talks about replicating the full gamut of human intelligence any more. Instead we see a retreat into specialized subproblems."[15] The lofty goals of artificial intelligence had been abandoned in many quarters.

The observed gulf between promises and reality led to more than one "AI winter"—a seasonal metaphor constantly used for the drying up of lavish government funding. The 1970s and early 1980s saw a second boom-and-bust cycle, this time in expert systems. The builders of such systems aimed to collect information from human experts, organize that information into systematic procedures, and then implement those procedures on computers, to undertake tasks such as medical diagnosis. Despite limited success in a few domains and the silent integration in many everyday systems, the systems proved brittle and the market for them tanked as the 1980s came to a close.[16] The confidence of luminaries like Herbert Simon proved to be misplaced.

In the eyes of those seeking to reorient artificial intelligence, the entire project of rules-based artificial intelligence rested on a misconception about human knowledge: it wasn't articulable easily in simple rules. It wasn't bookish knowledge—it was more like a practiced skill. Escaping this misconception, however, enabled a research effort to study the activities of experts without attempting to understand how they themselves make judgments. An article in 1993 noted, "Rather than asking an expert for domain knowledge, a machine learning algorithm observes expert tasks and induces rules emulating expert decisions."[17] Unlike more ambitious forms of artificial intelligence seeking to emulate *how* humans make decisions, the makers of such algorithms viewed them as acting in no way like human brains. Rather than a focus on logic, or interviewing experts, machine learning of this sort focused even more centrally on data, data about humans and data partially classified by humans. And it did so using the tools of pattern recognition, statistics, and neural networks that the symbolic artificial intelligence community had in large part shunned. Much of this work happened at industrial labs with an engineering mindset, deep pockets, and access to large amounts of expensive computer time, places like Bell Labs and IBM.

Outside the US and UK

Data-driven computational statistics also developed in opposition to mathematical statistics and symbolic AI outside the Anglophone world. In France, Jean-Paul Benzécri created a powerful school of *"analyse des données"* [analysis of data] focused on more powerful exploratory and descriptive statistics using computers. "The progress of the 'analyse des données' due to computers," he wrote, "will not continue without upsetting all of statistics."[18] In Japan, Hayashi Chikiō developed a set of practices he named *Deta no Kagaku*, the "Science of Data," as an alternative to mathematical statistics, which he described as "good-for-nothing and not understandable."[19]

Developments in the Soviet Union were probably of the most consequence to the recent history of data and its analysis. In 2006, the machine learning specialist Vladimir Vapnik reflected on the transformations in computer learning in the USSR decades prior. Rejecting dominant statistical approaches, Vapnik and like-minded colleagues created "predictive (discriminative) models of induction." In such an approach, "predictive models do not necessarily connect prediction of an event with [an] understanding of the law that governs the event; they are just looking for a function that explains the data best."[20] Vapnik came to this instrumentalist approach, and the high-dimensional data sets, as a member of the Institute of Control Sciences of the Academy of Sciences of the USSR in the 1960s and 1970s.[21] While anti-Semitism directed toward him and his refusenik advisor likely stymied his career, being in the institute allowed Vapnik to participate in the flourishing of a highly computationally focused learning approach applied to large sets. Vapnik moved to the United States and worked at Bell Labs. In the US and USSR alike, pattern recognition and control theory researchers viewed themselves as distant from symbolic artificial intelligence and from classical academic statistics.

All these tendencies found financial and moral support at Bell Labs in the 1990s. At that time, Bell Labs had hired an impressive array of international researchers who pioneered new methods and branches of machine learning, including future luminaries Yann LeCun, Yoshua Bengio, Rich Sutton, Rob Schapire, and others. The technique most associated with Vapnik, called support-vector machines (SVMs), came to fruition in a remarkable collaboration there, where he joined forces with the French researcher Isabelle Guyon. Like other major examples of development in the computational data sciences, Vapnik worked under the imperative of contending with high-dimensional data within a funding regime supporting it, and without the burden of producing symbolic artificial intelligence.[22] For all its importance, Bell was not the only site for such dramatic possibilities. At IBM, Stanford professor Xiaochang Li has shown, a similar convergence of statistical knowledge, an engineering mindset, large sets of speech data, and access to computing power permitted a dramatic transformation of speech recognition.[23] These industrial sites presaged—and made possible—much larger developments.

The Subterranean World of Neural Nets

Despite the numerous strikes against neural nets, a passel of researchers from Japan to France continued to research neural networks, both for doing predictive machine learning and for learning more about animal brains. And, despite the hostility of many in the machine learning and AI communities to neural nets, Bell Labs and especially the Canadian organization CIFAR provided funding necessary to keep the research going, sustaining memories of the power of nets and their relatively recent successes at tasks like recognizing digits. The convoluted story at the intersection of computation and neuroscience is well told elsewhere, so we only sketch the key developments.[24] At a very general level, we can say that by the mid-1980s, several researchers hit upon similar ideas for how to train multilayer neural nets, through a process known as "backpropagation."[25] When the network incorrectly classifies something, say classifying an image of a hot dog as a dog, any error is used to change values deeper into the network, thus training the "neurons" to make fewer errors and make more correct decisions. This algorithm could in principle put to rest some of the reasons for the rejection of neural networks at the hands of the AI practitioners, as these "deep" networks could discriminate among far more complex things than the simpler networks of the 1960s. The development of parallel computers made this work seem more computationally plausible. A parallel computer involves a large number of processors working on the same problem rather than a single or small number of very powerful processors working individually.

As if the opposition of the old symbolic AI folks wasn't enough, many in the newly data-intensive machine learning community perceived neural networks as dated and wasteful, a throwback to earlier days, long since surpassed by better and cheaper algorithms. Unlike many of the best algorithms of the day, neural networks lacked certain important mathematical properties, to the dismay of many in the community. The new backpropagation algorithm was slow, computationally intensive, and provided no guarantee that the network had been trained to find the very best answer—a criterion central in the field of mathematical optimization and important to early generations of AI folks and many in more statistical communities. Even advocates of the new networks could not understand or explain in any detail why the networks made the predictions they did—they were truly black boxes. They worked well enough at prediction, but not through rules humans could understand in any ordinary way, and only at massive computational cost. The techniques found some early successes in industry, such as in the reading of numbers of bank checks, but carried little academic prestige well into the 2010s.

Despite the successes of the new forms of neural networks, a nearly Biblical period of exile for neural net researchers followed—at least in the eyes of the true believers. For the team at Bell Labs, the opposition—many of their best friends—was in the next room, filled with proponents of so-called "kernel" machines that at the time seemed the likely victors in the battle among different machine learning algorithms.[26] These kernel machines—pioneered by Vapnik—were powerfully predictive but they also had important mathematical qualities, beloved of mathematically inclined researchers, that neural nets lacked. "Until 2010," one anonymous French researcher noted, doing nets was "a has been thing." Other researchers had little interest, illustrated by the cold shoulder given Yann LeCun, now the chief AI scientist at Meta. "I remember, LeCun was in the lab as an invited professor and we had to make someone go to dinner with him. Nobody wanted to go."[27] Key people even left the field, albeit temporarily. Despite their apparent success, Yann LeCun and Léon Bottou for example, French researchers recruited to Bell Labs, turned their attention to creating better alternatives to compressing images.

Ironically, however, the exploding success of other highly instrumentalist predictive algorithms was clearing the way for neural networks to become acceptable. The ground was shifting about how to evaluate algorithmic systems, just as data from the internet and computational power were exploding. By the 1990s, a growing literature revealed, the statistician Leo Breiman argued that "combining a multiple set of predictors, all constructed using the same date [sic], can lead to dramatic decreases in test error." This predictive success came, however, at great cost: the models were increasingly inscrutable to human beings.[28] These amazing predictive techniques did not render anything like rules intelligible to human beings. Neural nets, however, were always inscrutable, whereas a fundamental virtue of many other machine learning algorithms had been that they were interpretable. The predictive gains from these new ensemble models and neural nets were widely seen as massive, and to an increasing number of practitioners among different disciplines, have come to overshadow the massive opacity of the predictive ensemble generated. Bell Labs researchers produced much of the work celebrating ensemble modeling, and took part as the winners in contests of algorithmic prediction, most famously the Netflix Prize, awarded in 2009. Increasingly, machine learning practitioners were abandoning using any one family of predictive models in favor of combining many different predictors. This dramatic success of ensembles amplified the ethic of prediction over interpretation. Predictive capacity, more than any other virtue, became ever more ascendant in machine learning. Demands for interpretability, demands for rules understandable to human beings, were fading.

In precisely this context could neural networks return from their exile, in particular large, many-layer neural networks now rebranded as "deep learning." Even in the 1980s, the inscrutability of neural nets made them problematic, if not suspect; the renaissance of neural networks from around 2012 rests squarely on the legitimation of such ensemble models, for commerce, for spies, and for science.[29]

In 2012 a neural network dramatically outperformed all other contenders in an annual competition to predict the proper descriptive labels for objects found in ImageNet, a large data set of images assembled by Stanford professor Fei-Fei Li and her teams.[30] By the next year, all the major competitors had abandoned other families of algorithms in favor of their own versions of neural nets.[31] Supporters of neural nets—long shunned in the research community—felt vindicated. Decades of mockery of their approach suddenly seemed misguided, and journalists and academics alike began telling tales of their heroic return from unjustified exile.

The entire competition rested on a vast abundance of unseen labor, through the use of Amazon's Mechanical Turk, a crowdsourcing marketplace launched in 2005 that permits anyone to hire large numbers of remote workers to perform tasks—typically tasks computers can't yet do automatically. The human classification of these images was the largest use of Amazon Mechanical Turk distributed labor up to that point, and thus also rested on the abundant funding needed to pay for that labor, estimated to include some 25,000 people by 2010.[32] Crowdsourced workers put 14 million images into over 21,000 categories. Their labor in classifying, right and wrong, provided the "ground truth" for the algorithmic models to try to predict based on these enormous—for the time—data sets.[33]

After their long time in the wilderness, with performance gains benefiting from massive computation along with such massive data sets, neural nets came to perform better than other approaches. Their success in 2012 is often portrayed as a dramatic break, but a more sober historical picture suggests otherwise. As "deep learning," neural nets had become acceptable, in many ways because other models had moved far away from straightforward algorithms computed in short of amounts of time. One strike against neural nets was that they were expensive to train because they required so much computer time and capacity. By 2010, all the competitors were similarly computationally expensive, requiring either long training times with one computer or multiple computers working in parallel—usually both. Performing extensive computational work required large amounts of money. A second strike was that neural nets might be good at—if slow to start—making predictions, but they offered little explanation of those predictions. But the same had become true of competing approaches. By 2012, competitors were using extremely complex congeries of algorithms bundled together into an ensemble that voted in making predictions: they were almost as complex as the neural nets. Like neural nets, large ensemble models, complicated kernel spaces, and other approaches had similarly come to privilege predictive power over human interpretability.

Deep learning ascended only after the philosophical and mathematical objections to algorithmic systems focused almost exclusively on prediction ceased to matter—in industry, in the military, and, to a lesser extent, in academia. Deep learning was widely understood to provide the best predictors, and thus the most successful approach *if one's goals were prediction*. With these predictive successes, and the narrowing of expectations of what an algorithmic system should provide, the defects of neural nets in the eyes of statisticians and computer scientists became easier to ignore. Even with the new techniques for training neural networks, doing so required enormous amounts of data, huge computational power, and deep pockets to provide electricity for that training at scale.

What people expected from machine learning models had decisively changed, and new variants of hardware, called graphics processing units (GPU), made training neural nets easier and faster.[34] Above all, the cold hard cash needed to train extremely large models was becoming increasingly available only to researchers through companies like Google and the GPU manufacturer NVIDIA. In the years since, the models have only gotten larger, trained on ever-increasing data sets, with spiraling costs of computing, both in dollars and in carbon dioxide emissions.[35]

The redefinition of machine learning as focused on prediction, large data sets, and big computers was already under way when neural nets came into prominence. Machine learning, especially machine learning using neural nets, was rebranded as AI by corporate consultants and marketers, sometimes to the discomfort of researchers. The sheer scale and costs of this sort of research dramatically altered academic and even start-up research in machine learning. Only a few firms have the data, money, and computing power for the leading edge of algorithmic models, and researchers have come increasingly to depend on them, if not to work for them. "University labs and startups that wanted to develop and study AI found themselves," AI Now Institute faculty director and ex-Googler Meredith Whittaker explains, "requiring access to costly cloud-compute environments operated by big tech firms and scrambling for access to data, a dynamic that has only intensified since 2012."[36] As she explores, the tools for doing machine learning have been accessible and increasingly easy to use, but often they depend utterly on a small number of extremely well-resourced firms. (Our course at Columbia, for example, makes use of Google's product Colab, which allows us to teach a broad array of machine learning and statistical techniques at the cost of acclimatizing our students to the use of Google tools.)

Optimizing for What?

In 2015, *Science—The New Yorker* of the scientific world—carried a piece where two major researchers, Michael Jordan and Tom Mitchell, laid out the state of affairs in artificial intelligence. Having a model learned from data, rather than by hard coding of rules, now dominated AI. "Many developers of AI systems now recognize that, for many applications, it can be far easier to train a system by showing it examples of desired input-output behavior than to program it manually by anticipating the desired response for all possible inputs."[37] The power—and applicability—of these algorithms came from a narrowing of the tasks to be performed. A machine learning system is successful in terms of some numerical way of representing what matters to you. The authors explain, "machine-learning algorithms can be viewed as searching through a large space of candidate programs, guided by training experience, to find a program that optimizes the performance metric."[38] In other words, a machine learning algorithm produces a large number of candidate programs for doing some task, say classifying dogs vs. cats, and searches for one that best does so, according to a metric that you designate in advance: accuracy, with the smallest number of false positives, for example. As Pat Langley had complained, "Machine learning focused initially on using and acquiring knowledge cast as rich relational structures, while many researchers now appear to care only about statistics."[39]

What was gained? And what was lost? Prediction prevailed—prevailed over a modeling of the underlying processes of the thing being predicted. And it prevailed over a concern with being able to interpret and understand the processes of the algorithm in making those predictions. Neural networks were long anathema in part because they were opaque. But when most of the other algorithms had become similarly opaque, and the fundamental goals were predictive, the faults of neural networks no longer mattered as they once had.[40]

These dramatic changes in machine learning enabled and were funded by its explosion within the corporate sector. There the metrics involve money, at least indirectly: page views, online purchases, time spent on a social network, "engagement."

Netflix Prize

"We are quite curious, really," Netflix announced, to "the tune of one million dollars." In 2006, Netflix offered this substantial purse to anyone who could dramatically improve its algorithm for recommending movies to users:

> we thought we'd make a contest out of finding the answer. It's "easy" really. We provide you with a lot of anonymous rating data, and a prediction accuracy bar that is 10% better than what [Netflix's algorithm] can do on the same training data set. . . . If you develop a system that we judge most beats that bar on the qualifying test set we provide, you get serious money and the bragging rights.

Competitors had to make their algorithms public:

> But (and you knew there would be a catch, right?) only if you share your method with us and describe to the world how you did it and why it works.[41]

The company made a substantial data set available: ratings on 17,770 movies and 480,189 anonymous users over approximately seven years, for a total of 100,480,507 ratings. While such large data sets were increasing the currency of large internet firms, researchers rarely had access. Bell Labs' Chris Volinksy explained that Netflix "made a brilliant move by realizing that there was a research community out there that worked on these kinds of models and was starving for data."[42] Machine learning was different—more powerful— with very large data sets. But they remained rare.

In 2009, the team BellKor's Pragmatic Chaos won the million dollars for building a superior movie recommender system, beating out its competition, The Ensemble, by twenty minutes. The winning team's name brings together the names of four separate groups who joined their efforts.

Their social combination was mirrored in their winning

algorithm, which combined the efforts of all four groups into a massive predictive ensemble, with models from all parts of machine learning brought together. Lacking constraints of intelligibility or explainability, the single performance metric enabled a peculiar social coordination organized through email and discussion boards: a competitive, community-organizing task, the so-called "common task framework." The data scientist David Donoho calls the competitive focus on a common score to be maximized the "secret sauce" behind the transformative success of machine learning on large sets of data in the past twenty years. The common task allows "a total focus on optimization of empirical performance, which . . . allows large numbers of researchers to compete at any given common task challenge, and allows for efficient and unemotional judging of challenge winners."[43] The common task framework, Donoho argues further, "leads immediately to applications in a real-world application. In the process of winning a competition, a prediction rule has necessarily been tested, and so is essentially ready for immediate deployment."[44]

Indeed, deploying machine learning often involves algorithmically maximizing a quantified value. In industry, such a quantitative goal is termed a "key performance indicator"; a numerical measure correlated either with business goals, or with product goals such as page views, time spent on an article or video, or "engagement" more generally—or ideally with both!

At the close of the Netflix competition, an MIT fellow, Michael Schrage, explained, "The great advantage of the prize model is that it moves work away from the realm of the beauty contest to being performance-oriented." Such celebration naturally rests on believing in the superiority of some metric: "It's the results produced that matters."[45] To say something matters or doesn't is a statement of values.

Rather than valuing complex phenomena like beauty, proponents of machine learning largely valued phenomena capable of being given a quantified measure. Early in this book, we quoted testy Germans upset at the new "vulgar" (quantitative) statisticians, who confused numbers for knowledge of a land or a people, and misunderstood value. Machine learning, as it developed by 2000, was poised to be an apotheosis of the numerically focused statistician—powerful just because it was limited in purview. The ethical and political concerns of contemporary AI circle around the recasting of AI as the optimization of metrics.

The Netflix competition illustrates how the machine learning approach came to be used in the 1990s and 2000s far beyond academic centers and industrial research laboratories, on an array of commercial, industrial, medical, policing, and military applications, at once dizzying, exciting, controversial, and sometimes discriminatory. Those advocating for industrial-scale machine learning and building it into business and governmental practices came to be known as "data scientists" by the 2010s. While producing tools enabling everyone from scientists to journalists to draw upon machine learning, they simultaneously drew on a broad array of other skills to scale machine learning, to make it figure centrally in the infrastructures mediating our communications, our science, our news, and our politics.

CHAPTER 10

The Science of Data

the altered field will be called "data science" . . . technical areas of
data science should be judged by the extent to which they enable
the analyst to learn from data.
> —Bell Labs statistician William Cleveland, "Data Science:
> An Action Plan for Expanding the Technical Areas
> of the Field of Statistics," 2001

At Facebook we felt like different job titles like research scientist,
business analyst didn't quite cut it for the diversity of things that
you might do in my group. A "data scientist" might build a mul-
tistage processing pipeline in Python, design a hypothesis test,
perform a regression analysis over data samples with R, design
and implement an algorithm in Hadoop, or communicate the
results of our analyses to other members of the organization in
a clear and concise fashion. To capture this, we came up with the
title "data scientist."
> —Jeff Hammerbacher, "Information Platforms
> and the Rise of the Data Scientist," 2009

"I saw the best minds of my generation destroyed by mad-
ness," wrote the poet Allen Ginsberg. In clause after
clause, Ginsberg sang of the gulf between higher aspiration
and the realities of Cold War America: "angelheaded hipsters
burning for the ancient heavenly connection to the starry
dynamo in the machinery of night"—and the chasm experi-
enced by students with the increasingly militarized universities: "who passed through universities with radiant cool

eyes hallucinating Arkansas and Blake-light tragedy among the scholars of war."[1] In 2011, Jeff Hammerbacher, a former Facebook data team leader, riffing on Ginsberg, bemoaned, "The best minds of my generation are thinking about how to make people click ads. That sucks."[2] Of all the things to optimize, a generation had chosen manipulating attention. This chapter traces the evolution of "data science," a term which gained prominence first at companies making people click on ads, but whose history stretches from the Cold War to the present.

Along with DJ Patil, Hammerbacher is credited with coining the term "data scientist" to describe a crucial new role in the corporate world from start-ups to Fortune 500 corporations. What does a data scientist do differently than practitioners of all the various quantitative approaches to the world we've seen? What exactly is "data science"? Definitions, we will see, vary. Industrial data science came to mean machine learning and statistics combined with the software engineering and concrete data work needed to build digital products and services. In academic research, the term is capacious, extending beyond statistics to include the broader and less "technical" skills needed for making sense of the world through data, from the messiness of "data janitorial work" to the nuances of communicating results through data. Rather than abstractly "burning for the ancient heavenly connection," the term speaks to the practical complexities of such work, starting with data analysis getting grubby with data. Riffing on another, very different, Cold War writer, the data scientist Joel Grus satirized the expectation a "data scientist" had mastered the wide diversity of data tasks needed in industry:

> a data scientist should be able to run a regression, write a sql query, scrape a web site, design an experiment, factor matrices, use a data frame, pretend to

understand deep learning, steal from the d3 gallery,
argue r versus python, think in mapreduce, update
a prior, build a dashboard, clean up messy data, test
a hypothesis, talk to a businessperson, script a shell,
code on a whiteboard, hack a p-value, machine-learn
a model. specialization is for engineers.*

As the field rose to prominence in industry and academia, with associated job opportunities, funding opportunities, and new departments and degrees, employers and administrators sought to define things more precisely. Often, trying to nail down "data science" devolves into a verbal tussle in the online comment sections which coevolved with the internet. Rather than insist on one definition of "data science," we seek to outline contours of contestation around the term. For a decade now, in presentations, through memes, in comments to posts, practitioners have fought over what the term really stands for, in contrast to say statistics, machine learning, or earlier "data mining." The arguments fundamentally concern who has authority and who gains capacities to rearrange power in dealing with data. And they concern who ultimately gets the funding—in corporations, in academia, and from the government.

To be clear, there was good reason for excitement and funding. In a variety of industries, making sense of

---

* Grus, "The Road to Data Science." He references Robert Heinlein: "A human being should be able to change a diaper, plan an invasion, butcher a hog, conn a ship, design a building, write a sonnet, balance accounts, build a wall, set a bone, comfort the dying, take orders, give orders, cooperate, act alone, solve equations, analyze a new problem, pitch manure, program a computer, cook a tasty meal, fight efficiently, die gallantly. Specialization is for insects." Robert A. Heinlein, *Time Enough for Love: The Lives of Lazarus Long; a Novel* (New York: Putnam, 1973).

the world through data had been transformational. The ability to recommend the right product and content to commercial users made possible a so-called "long tail" business model.[3] Similarly, in commercial software, we've become used to phones as devices we can talk "to," not "on," as speech recognition has improved through multiple quantum leaps. In finance, the single most profitable fund, the Medallion Fund at Renaissance Technologies, trades using statistical analysis, along with considerable attention to the software engineering needed to gather data, learn models, and execute trades.[4] In biology and human health, it was quickly realized that the sequencing of whole genomes in the 1990s had the potential to change our understanding of complex human diseases through data. "Biology is in the midst of an intellectual and experimental sea change," declared the biologist Shirley Tilghman in the first sentence of an article in *Nature* in 2000. "Essentially the discipline is moving from being largely a data-poor science to becoming a data-rich science." In a wide variety of fields of human endeavor, it was clear that "new technology permitted entirely new questions," that "will require . . . new sets of analytical tools."[5]

Merely Statistics or Not . . .

In 2011, a mathematician turned data guru, Cathy O'Neil, and a statistician, Cosma Shalizi, got into a good-natured tiff on the internet about the nature of the "sexiest" career of the moment: the data scientist. O'Neil argued that much of data science involves getting to the point one can use statistics:

> In other words, once we boil something down to a
> question in statistics it's kind of a breeze. Even so,

nothing is ever as standard as you would actually find in a stats class—the chances of being asked a question similar to a stats class is zero.

Data scientists had a much wider range of questions asked about less standardized data. And thus, they needed different competencies.

I would add that it's really not about familiarity with a specific set of tools that defines a data scientist. Rather, it's about being a craftsperson (and a salesman) with those tools.

To set up an analogy: I'm not a chef because I know about casserole dishes.[6]

Shalizi demurred:

What strikes me about it, though, is that the skills she's describing a good "data scientist" as having are a subset of the skills of a good statistician. At most, they are a subset of the skills of a good *computationally competent* statistician.[7]

At the heart of the difference between industrial data science and the academic worlds of statistics and machine learning is the prioritizing and celebration of capabilities for dealing with problematic real-world data, often in large infrastructures. "Dealing with" stands for the craft of getting it into forms that standard algorithms can use. But it often also means wrangling very large data sets into distributed databases capable of contending with them. And unlike those academic fields data science is often understood as fundamentally oriented by the business needs of an organization, be it a corporation or government agency.

MANY ACADEMIC STATISTICIANS and machine learners have disparaged these more craft-like elements as lower on the knowledge scale and perhaps even simple to learn. Though less theoretical subjects are not necessarily easy to master, they are not enough; as O'Neil quipped, knowledge of casserole dishes does not a chef make.

At its most hubristic, data science is presented as a master discipline, capable of reorienting the sciences, the commercial world, and governance itself. It is a candidate for reorganizing knowledge and power, now built into institutions dominating much of our lives.

The roots of data science are gnarled; they include rarified math but also much engineering craft; university lecture halls but also in marketing departments and the war rooms of politicians. The impure blended quality of data science speaks to a key narrative we have been tracing: the coming together of increasingly automatic forms of decision-making with large-scale infrastructures enabling those processes. Data science arises from a coming together of statistics working with real-world data, machine learning, and analytical processing of data within businesses large and small. The story requires moving between the world of computational statisticians warning about "over-mathematization" and developments within industry. We begin with some heretical statisticians who, informed by their real-world experiences, exhorted their academic field to move closer to the data.

"Data Analysis," 1960s–1990s

In 1974 the Princeton–Bell Labs mathematician John Tukey agreed to speak at the National Security Agency about "exploratory data analysis," asking that the agency provide "2 screens and 2 projectors for large transparencies."[8] Long a scientific advisor to the NSA following his involvement with cryptography during World War II, Tukey had, since the 1940s, been creating new tools for exploring data, large and small, using all manner of statistical and graphical methods. Initially focused on paper tools for exploring data, he was at the forefront of the move to computers for graphing and analyzing data. Twenty-five years before, NSA's Kullback had invited Tukey to a "symposium on the general problem of data storage and retrieval"—based in part on Tukey's recommendation that the NSA look into the problem.[9] The symposium was to consider what were the data storage and retrieval issues in general—and what were those particularly to the NSA.[10]

Less important than the still-classified work Tukey did were the attitudes toward statistics and data he encouraged within the NSA—and within the unclassified world. Tukey worked for decades to transform the practical statistical work on large data sets of the war into far more general use toolsets—and mindsets. In his career, he worked on everything from the census to missiles. The tools whose creation he encouraged and the graphical techniques he advocated such as the box-plot saturate contemporary data practices, including middle school standardized exams.

Informed by the large-scale data analysis needed during World War II, Tukey provided a programmatic statement of a changed approach to data and sought to make tools to realize it. In a 1962 manifesto, Tukey called for a new approach he dubbed "data analysis" that would be dedicated as much to discovery as to confirmation:

> Data analysis, and the parts of statistics which
> adhere to it, must then take on the characteris-

tics of a science rather than those of mathematics, specifically:

1. Data analysis must seek for scope and usefulness rather than security.
2. Data analysis must be willing to err moderately often in order that inadequate evidence shall more often *suggest* the right answer.
3. Data analysis must use mathematical argument and mathematical results as bases for judgment rather than as bases for proofs or stamps of validity.[11]

As a scientific practice, Tukey described data analysis as an art, not a logically closed discipline. Tukey was crystallizing an alternate approach to academic statistics, one that used the mathematical power of statistical thinking for exploratory as well as confirmatory purposes, and one that might be applicable to observational data, not exclusively to data produced as part of an experimental trial. Thanks to the support of the mathematician Mina Rees and the efforts of the statistician Harold Hotelling and others, as we saw above, the great successes of highly *applied* statistics during World War II were channeled into financial and symbolic support for the creation of a mathematically focused, theoretical statistics in the United States and in Europe, rather than a more practically oriented, data-focused statistics. Before long, in the eyes of critics such as Tukey, practical data collection and analysis had been sacrificed at the altar of mathematical sophistication and rigor. He was pushing against the dominant tendency of statisticians in universities to force statistics to emulate the abstract form of pure mathematics as much as possible, a position that, in his eyes, involved too much rigor and not enough working with data. Recall that the statistician Hotelling, in contrast, worried about the corrupting influence of young students being exposed to *too much actual data*. (To be clear, Tukey was no stranger to mathematical rigor, having completed his PhD in topology, a branch of pure mathematics, the same year World War II began.)

Tukey pursued this data analysis as a key member of Bell Labs, drawing on his wartime experience and decades-long work for the NSA and the military services. Thanks to "war problems" in the 1940s, Tukey explained in an interview, "it was natural to regard statistics as something that had the purpose of being used on data—maybe not directly, but at most at some remove. Now, I can't believe that other people who had practical experience failed to have this view, but they certainly—I would say—failed to advertise it."[12] In the 1960s and 1970s, Tukey and other critics complained that relatively few within academic mathematical statistics and its allied branches, such as econometrics, celebrated the practical cultivation of data analysis and forms of judgment as a central endeavor. As we've seen in the previous chapter, data analysis in the form of pattern recognition flourished elsewhere, in the penumbra of mathematical statistics and other well-established disciplines, in corporate research labs and engineering departments, under various names.

In the atmosphere of Bell Labs, Tukey and his collaborators created a wide variety of statistical and computational tools needed to make data analysis a reality. Sixteen years later, in a practical textbook, he explained, "exploratory data analysis" (EDA) is "detective work—numerical detective work—or counting detective work—or graphical detective work." EDA offered some "general understandings" useful across domains of detective work. "The processes of criminal justice are clearly divided between the search for the evidence—in Anglo-Saxon lands the responsibility of the police and other investigative forces—and the evaluation of the evidence's strength—a matter for juries and judges. In data analysis a similar distinction is helpful. Exploratory data analysis is detective in character."[13] Exploratory data analysis is a technical craft—and Tukey celebrated the creation of new tools for that craft.

Tukey's 1978 textbook, whose draft had circulated for years in Bell Labs circles and beyond, offered a survey of the arts of exploring data through potent means of "re-expression." **"We have not,"** he explained in bold type, **"looked at our results until we have displayed them effectively."**[14] Effective display means developing proficiency with many forms of visualizing data, Tukey emphasized, "much more creative effort is needed to pictorialize the output from data analysis. . . . For humans, the use of appropriate pictures offers the possibility of great flexibility all along the scale from broad summary to fine detail, since pictures can be viewed in so many ways." While Tukey forecast that computers would soon predominate in graphing, in the meanwhile he had developed a variety of practices for visualizing data by hand.

Colleagues at Bell Labs brilliantly continued work along the lines Tukey had set out and beyond, now increasingly in the context of the explosion of data in commercial and scientific systems. In 1993, his Bell Labs colleague John Chambers penned his own updated manifesto calling for the expansion of the ambitions of statistics. Capturing the difference between one who merely knows casserole dishes and a chef, Chambers contrasted *lesser statistics*, "as defined by texts, journals, and doctoral dissertations," from *greater statistics*, "inclusive, eclectic with respect to methodology, closely associated with other disciplines, and practiced by many outside of academia and often outside professional statistics."[15] Unlike lesser statistics, greater statistics concerns itself not only with simplified, clean data, and not only with academic publication:

Three broad categories characterize work in greater statistics:

— *preparing* data, including planning, collection, organization, and validation

— *analyzing* data, by models or other summaries

— *presenting* data in written, graphical or other form[16]

Preparing and presenting in real-world situations, Chambers insisted, was "rich with intellectual challenges as well as practical importance." Tukey worked with the NSA to deal with vast data. Chambers noted the challenge presented by the increasing accumulation of data in real world systems. "Many mundane . . . activities generate large quantities of potentially valuable data. Examples . . . include retail sales, billing, and inventory management. The data were not generated for the purpose of learning; however, the potential for learning is great."[17] Chambers was writing in the context of Bell Labs, with access to the nation's telecommunications data as well as the diverse data encountered by Bell Labs researchers in their partnerships with the US government.

A similar observation—that large data sets gathered for one purpose may yield potential new kinds of scientific and commercial knowledge—would be made in a diversity of computational fields over the coming decades. Financial data and their practical analysis would give rise to technical analysis, statistical arbitrage, and later, with more computational engineering, the field of high frequency trading. Similarly, computational biology in the 1990s and 2000s exploded with analysis of differing genomes as well as high-throughput biological assays for understanding genetic networks, large-scale mining of electronic health records, and clinical informatics.[18] In industry, applied, computational statistical methods changed the way companies recommended books and movies early in the rise of e-commerce, then later the same techniques would be applied to wine, shoes, and eventually information and communication. Each of these fields had its own "data moment" as it discovered anew how large quantities of data, generated for purposes other than learning, could be valuable given a bit of statistical analysis surrounded by an infrastructure need to gather, process, and productize insights from these data. Chambers, Tukey, and others argued that the statistical analysis was a mere part of this project—the mathematical nugget at the core of "greater" statistics. But they were also warning that academic statistics was doomed to irrelevance if it didn't begin providing the tools for learning from this data.

In 1998, Chambers received the Software System Award of the Association for Computing Machinery for the S system for data analysis and graphics presentation, "which has forever altered how people analyze, visualize, and manipulate data."[19] An open-source language R based on S became a dominant platform for computationally orientated statisticians and especially for work in graphical analysis and presentation.

The Bell Labs crowd thus created tool sets and attitudes to enable data analysis, both of traditional statistical forms and of a broader approach to statistics. They also celebrated the power of graphical methods.

A few years later, another Bell Labs statistician, William Cleveland, explicitly called for creating the field of "data science," a radical overhauling of statistics around its utility for practical data analysis. Statisticians had much to offer computer scientists—and computer scientists had much to teach statisticians: "the knowledge among computer scientists about how to think of and approach the analysis of data is limited, just as the knowledge of computing environments by statisticians is limited. A merger of the knowledge bases would produce a powerful force for innovation. This suggests that statisticians should look to computing for knowledge today, just as data science looked to mathematics in the past."[20] Universities needed to change.

The heretical statisticians of Bell Labs—Tukey, Chambers, Cleveland, and others—were not the only ones in the late twentieth century to see that there was a new field to be created by applying statistics to massive data sets. This realization was also brewing among those in industry and in academia trying to create the technology needed to store, secure, and search the large data sets increasingly produced in research, commercial, and government settings. By the 1990s, many in the community working on what they called "very large databases" were worrying about the lack of technologies capable of analyzing the data produced through everyday online and offline transactions.[21] Answering this challenge required new tech, new attitudes, and the definition and empowerment of a new kind of practitioner.

Data Mining, Early 1990s

By the late 1980s, the tools for analyzing and learning from the rapid expanding stores of business data were widely seen as increasingly inadequate, as we saw at the end of chapter 8. Similar stories held true with scientific, military, and intelligence data. In 1998, amid the blossoming of large-scale corporate, government, and academic "data warehouses," then–Microsoft researcher Usama Fayyad explained:

If I were to draw on a historical analogy of where we stand today with regards to digital information manipulation, navigation, and exploitation, I find myself thinking of Ancient Egypt. . . . A large data store today, in practice, is not very far from being a grand, write-only, data tomb.[22]

Much of the interesting big data is big in two different ways: it involves observations about say, a large number of people or a large number of purchases; and it involves, for each one of those observations, a large number of variables. The last point—called the "high-dimensionality" of data—has a major mathematical challenge that accompanies it. As the number of dimensions gets larger, the mathematical techniques used for comparing data points become problematic, and the amount of data necessary to achieve higher levels of confidence about conclusions becomes larger. Corporate, military, and intelligence data required the means for contending with high dimensionality in real time.

A movement known as "data mining" emerged in the early 1990s to leverage the growing untapped stores of corporate and scientific data. Data mining, or, as it more formally was branded, Knowledge Discovery in Databases (KDD), is the activity of creating nontrivial knowledge suitable for action from databases of huge size and dimensionality.[23] Data mining focused on databases of very large size—millions or billions of records, often with every record typically including a large number of elements. For each record in a retail database, a data mining operation might seek unexpected relationships among the item purchased, the store's zip code, the purchaser's zip code, variety of credit card, time of day, date of birth, other items purchased at the same time, even every item viewed, or the history of every previous item purchased or returned. Performing reasonably fast analyses of high-dimensional, messy real-world data is central to the identity and purpose of data mining, even more so than in pattern recognition or academic machine learning. Sophisticated statistical and machine learning algorithms before the 1990s were typically devised for sets of data that can easily fit in memory, or that require a relatively small use of slower disk access. Adapting such algorithms to huge quantities of data that cannot be held in memory proved far from obvious.

It's not just applying statistics or machine learning to a bigger problem. Key developments in data mining involve efforts to choose among the trade-offs necessary to make algorithms scale.[24] The ability to contend with scale would, in turn, dramatically reshape machine learning in practice. Data miners drew heavily upon machine learning, and they did much to make it scale in new ways. In the late 1980s and up into the present a pattern begins to emerge: algorithmic advocates take up a particular algorithm, offer a series of suggested improvements, often as part of doctoral study, and then become advocates for versions of those algorithms across various scientific and industrial domains. From a wild west of potential algorithms, a small set of powerful approaches to learning from data emerged as the most prized algorithms in data mining and machine learning.

Through data unwedded to theory, data mining promised to overcome usual ways of dividing and understanding the world. Statisticians working in the mold of Fisher would ask, "Are higher-income people prone to be more loyal to a warehouse club than those with lower income levels?" and test the hypothesis. "Data mining, on the other hand, potentially would provide more insight by pointing out other factors contributing to store loyalty that the analyst would not otherwise have been able to consider testing."[25] Scientists intrigued by the potential of these approaches made similar claims. In 1999 Patrick Brown and David Botstein explained, "Exploration means looking around, observing, describing, and mapping undiscovered territory, not testing theories or models. The goal is to discover things we neither knew nor expected."[26]

In the late 1990s, IBM's Almaden Labs in San José hosted an ongoing seminar series that brought in academic researchers, industrial researchers, and IBM's own employees, and more generally served as a center of sociability for the local data mining community.[27] Many of the papers presented there would become standard transformative works in scaling statistical and machine learning algorithms for use in existing hardware with large data sets.

One Wednesday morning in November 1997, a Stanford computer science graduate student came down to Almaden to speak on the topic of "Mining the Web." He explained:

> A new project at Stanford is the WebBase project. The goals are to collect a large amount of data from the Web and to make it available for research. While the project is relatively new (several months), it has already produced some interesting results.

The speaker, Sergey Brin, was the organizing force of a data mining group, MIDAS (Mining Data At Stanford), with the support of several faculty members, each a pioneer in database management. At its regular meeting, the MIDAS group discussed the state of the field, from algorithms to ethics: "Topics range from admistrative [sic] issues and grant proposals to conference-style presentations by students and visitors."[28] For his talk at IBM, Brin promised to range widely over work he and others at Stanford were doing to contend with the vastness of the then still novel World Wide Web.

> I will talk about some of the things we have discovered with this data and some algorithms that have been developed including link analysis, quality filtering, searching and phrase detection. [29]

The project would soon bear much algorithmic fruit. And, before long, many billions of dollars. The web page for MIDAS noted, "The most impressive and useful demo" of the group "is the super search engine, called Google, built by Larry Page and Sergey Brin."[30]

What Can You Do with a Web in Your Pocket: Late 1990s

Numerous computer science communities found themselves underprepared in the 1990s to contend with the vast, expanding, and decidedly non-curated World Wide Web. Older search and indexing tools were designed for highly standardized, curated, centralized collections of text or other data such as a collection of periodicals with their metadata. Researchers struggled with both the nonstandard and unstructured quality of web pages and their number.[31] Like many machine-learning algorithms, algorithms in the hoary field of information retrieval did not scale easily to the number of pages in the web. By the mid-1990s, search seemed to many an unpromising approach to the web. Major industry players focused increasingly on curated portals, exemplified by the approach of Yahoo. Search came to dominate after 2000, with the gradual, then exponential, rise of Google. And their approach to search emerged precisely from the concerns of the highly applied machine learning at the heart of data mining.

In 1998, Brin, in the Database group at Stanford, and his fellow graduate student Larry Page, in the Human-Computer Interaction group, drew on one of the most famous problems of the data miners—figuring out what items tend to go together when someone shops—called the "market basket" problem. Taking cues from looking at items in consumers' baskets at huge scale, they looked for associations within documents on the web. Their approach, called "dynamic data mining," did not "exhaustively explore the space of all possible association rules"—as the web was far too big to do so:

> when standard market basket data analysis is applied to data sets other than market baskets, producing useful output in a reasonable amount of time

is very difficult. For example, consider a data size with tens of millions of items and an average of 200 items per basket. . . . A traditional algorithm could not compute the large itemsets in the lifetime of the universe.[32]

Just as machine learning algorithms had to change to deal with the scale of early database mining, association mining algorithms had to change to deal with the scale of the early World Wide Web. In their adaptation of such an approach to commercial databases, Brin and Page exemplified the drive of practitioners focused on real-world databases to minimize disk and memory usage.

Brin and Page, along with their other collaborators, argued that the scale of the web, which made it so challenging, simultaneously made it deeply promising:

we take advantage of one central idea: the Web provides its own metadata. . . . This is because a substantial portion *of* the Web is *about* the Web. . . simple techniques that focus on a small subset of the potentially useful data can succeed due to the scale of the web.[33]

Based within a database community deeply interested in transforming existing statistical and machine learning techniques, Brin and his collaborators were prepared not just to deal with scale, but to make it into a central resource for discovery. Fundamentally, they realized that the scale of the web included vast human effort to classify and categorize the web in billions of piecemeal ways. Rather than creating any form of artificial intelligence capable of classifying the web itself by writing rules, they created a mechanism for leveraging human judgment at great scale.

Brin and Page's greatest breakthrough in mining the web came in adapting an everyday academic practice into algorithmic form most fruitful at vast scales. Following an insight of Page's, they adapted the idea of counting high-quality citations to gauge the authority or value of academic work. Web pages could be "ranked" as more or less authoritative by counting citations, that is, links to pages. More authoritative pages are those that have been linked to by other authoritative pages. The total number of links to a page counted far less than the authority of the pages linking to that page. They called the result PageRank, and they soon made it central to a new search engine, Google. Google search emerged from within a culture which fused database values about scaling to deal with data and practice with, later, the values of the machine learning community. Brin and Page recognized from the start the need for structuring databases capable of implementing the beautiful mathematics on fallible and limited machines. "Google's data structures are optimized so that a large document collection can be crawled, indexed, and searched with little cost."[34] A process for leveraging human judgment at mass scale, PageRank had to be materialized in a creatively designed set of databases. PageRank and its instantiation within commodity hardware in time led to the development of new architectures for distributed databases and distributed analytic processing, called BigTable and MapReduce respectively. The developments of these technologies for working with extremely large data sets figure centrally in the subsequent development of data science, as we will see. They made advanced machine learning at scale into technologies many users could deploy—if they had the right resources.

Military and intelligence concerns were never far from much of this work. In 2004, the National Security Agency and the Office of Naval Research sponsored a workshop on the analysis of "massive data streams." In the wake of 9/11, the intelligence and defense worlds needed the fruits of the data-centric enterprises they had long secretly cultivated. The chief of the Mathematics Research Group explained how much the NSA was profiting from data mining:

> We really have had some dramatic successes in terms of techniques we didn't have a year ago for looking for patterns in massive data, drawing conclusions and taking some known attributes of a situation and mining through the data to find new ones, and very algorithmic based, and really providing tools for our analysts. . . . For us, it is all about teaching the machines how to work for us, and teaching the machines is teaching the algorithms.[35]

In the 2000s, the NSA was not alone in profiting from this data-focused computational work; it was quickly imported into marketing, medicine, physics, education, criminal sentencing, social networking, and drone targeting. The challenges required to scale machine learning in the arenas of commercial, intelligence, and military data promoted the creation of technologies and technologists capable of dealing with ever-larger data sets.

From Data Mining to Big Data, 2000–2010

Although used from time to time before, the term "data scientist" flourished when it appeared as a job title at the internet platforms Facebook and LinkedIn. Far distant from internecine academic battles, these firms, like their rivals Google and Amazon, were accumulating data from everyday transactions on and off the web at an ever-increasing rate, probably rivaled only by the NSA itself. Storing, presenting, and analyzing this tremendous volume of data entailed staggering technical and intellectual challenges, challenges radically different in scale from analyzing smaller data sets on a desktop computer. The skills, practices, and software of statisticians would only be needed far later, once the challenge of scale had been met.

Data was accumulating fast as a succession of internet companies recorded as much as they could about their users, and, with the rise in the prominence of the advertising model among such information platform companies, about their corporate clients. Around the same time, the NSA received new authorities allowing it to capture untold amounts of internet and telephony traffic that overwhelmed its analytic capacities. Databases and the ability to analyze them were collapsing. Time and again the software and hardware couldn't handle the stream.

For example, when a key Facebook database approached a terabyte of data, Jeff Hammerbacher explained, the querying system "came to a sudden halt." It took three days for it to come back. Eventually Facebook adopted Hadoop, a powerful open-source framework for storing and analyzing large amounts of data; in large part developed by Yahoo, this technology allowed data to be stored over hundreds of servers and allowed analysis, based on a Google process known as MapReduce, to be divided among those many servers. Hadoop also allowed for a mix of "structured" and "unstructured data"—think of an address, with clearly demarcated blocks (name, street, zip code) versus the unbroken flow of text of a letter.

Similar stories ramified across industries old and new—and within academic precincts where new troves of data, especially generic data, overwhelmed old modes of computational analysis.

If too much data was the problem, it also offered great

opportunity. Three Google researchers celebrated what they called the "unreasonable effectiveness of data." They argued tons of data with simple models would almost always do better than complex models with little data.[36] Facebook and Google worked to leverage this new approach—as did the NSA.

A 1996 interview in the highly classified house magazine of the NSA turned to the question of the volume of world communications to be spied upon:

> Let me add to all of that the third biggest challenge facing us, and that is volume. And I could just end the sentence there and everything is said.[37]

By 2006, a top-secret email "Volume is our Friend" suggests a newfound confidence in the NSA's ability to contend with data overload: indeed, the enabling quality is central to the celebration of big data elsewhere. The bigger the volume, the better.

The resources dedicated to the Global War on Terror allowed the NSA to acquire enormous collection and analytical capacity—but it ever needed more. In 2008, NSA turned to the databases for big data produced within the open-source community and based on ideas from Google. Based on some ideas central to Google called BigTable, a group of scientists and programming within the NSA created—and released to the open-source community—a distributed database platform designed to accommodate graphs with billions of points requiring petabytes of storage capacity.[38] The War on Terror saw millions if not billions of US federal dollars pour into machine learning, computational statistics, and distributed computing. The intelligence agencies and military branches drew heavily upon academic and commercial developments, while providing a consistent stream of funding in all of the key fields.

An NSA job posting from around 2013 calls for a SIGINT "Informatist":

> hybrid computer scientist, analyst with work spanning the divide between the process-focused technical work and the content-focused analytical work.
> Responsibilities:
> - Combine information about the structure, syntax, and processing of data with the functions of gathering, organizing, and manipulating datasets in order to synthesize responses to customer information needs.
> - Apply scientific techniques to data evaluation, performing statistical inference and data mining.
> - Document and present the data analysis and its conclusions for assessment by full-performance analysts, developers, and their managers.[39]

In the secret world and in the corporate world alike, the new role of data scientist came to ever greater prominence. As in industry and academia, the NSA embraced a shift in the forms of excellence and knowledge that enterprises celebrated and empowered. A short overview of the many changes in the NSA's institutional culture from the Cold War to the present explained how the agency had abandoned a culture of "perfectionism" for something radically different. Amid the "winning" of the late Cold War, "NSA Valued in the 1980s, Accuracy, Deep Knowledge, Thorough Expertise, Productivity and Reputation." In the asymmetric world of a dizzying array of potential enemies, in contrast, "NSA valued in the 2000s . . . Speed—getting it 80 percent right now could make all the difference in saving lives. (Of course, if

it were targeting information that would mean killing inno-
cents 20 percent of the time.)"⁴⁰ The analysis is incisive and
scary. It's one thing for Netflix to make bad recommendations, quite another to advance grounds for surveillance,
drone strikes, or worse.

Artificial Artificial Intelligence

In their *Data Feminism*, Catherine D'Ignazio and Lauren
Klein insist on a key principle: "The work of data science,
like all work in the world, is the work of many hands."⁴¹
While the internet makes the collection of data easy at hitherto unknown scales, it didn't make processing it independent of human beings. Rather than eliminating human labor
and judgment, large-scale algorithm systems both displace
labor and fundamentally depend on other forms of labor.
Underlying all the new hardware and software, all the algorithms, was human work to make the data tractable. Some
of this labor fell within the purview of shiny new data scientists, but much if not most of the mucking fell to people
rarely visible in the workings of the corporations. Visions
of computers taking over all jobs are just false. "In order to
understand what automation does to human activity," insists
the scholar Antonio Casilli, "we must recognize and estimate
first the amount of work inscribed into automation itself."⁴²

Data labor is not new, and nor is its obfuscation in our
history: think of Bletchley Park or census workers in the
late nineteenth century. Yet the scale today is certainly
unprecedented—and enabled by the very systems in question.

From the start, Google's search algorithm leveraged the
implicit human ranking of web pages. To get the relatively
clean search results the company now delivers, it rests on
billions of human judgments to deem content explicit or not,
sexist or not, racist or not, as Sarah Roberts, Mary L. Gray,

Siddharth Suri, and Antonio Casilli have documented through detailed anthropological fieldwork and sociological study.[43] Building upon earlier work by Lucy Suchman and Shoshana Zuboff, these scholars all stress the processes of obfuscation of laborers worldwide, from India and the Philippines to the rural United States, an obfuscation that makes it appear the tech is doing the work—not the people. Gray and Suri explain:

> Billions of people consume website content, search engine queries, tweets, posts, and mobile-app-enabled services every day. They assume that their purchases are made possible by the magic of technology alone. But, in reality, they are being served by an international staff, quietly laboring in the background.[44]

Such labor makes the application of statistics and machine learning to large data sets possible. Casilli explains it: "At the antipodes of robotic fantasies sustaining the imagination of investors and media personalities are the myriad non-specialized click-workers performing the necessary work for choosing, improving, and making data interpretable."[45] Applying machine learning to the world requires data, even automatically collected data, to be made usable.

Critics have rightly noted that many supposed AI successes at some level involve continuing human decision-making, often at a vast scale. Amazon's Mechanical Turk, professor and former Googler Lilly Irani argues, "has allowed canonical AI projects to proceed by simulating AI's promise of computational intelligence with actual people."[46] But even far more automated systems depend on data classified and cleaned and produced by teams of laborers, usually far from the ludic environs of the software companies with their private chefs and foosball tables. Even as systems became superior at performing human-like tasks, they usually do so based on larger pools of human classified and produced data. "These workers power the tech industry," Irani explains further, "yet are out of sight and out of mind in the press and policy on diversifying the tech workplace. The diversity is there. It's just subcontracted and paid poorly." She continues: "These workers excel in doing what machines cannot. They have won the race against the machine, but they do not always even make minimum wage."[47] And this is true for data science, for good—and for bad.

Statistics Comes to Data Science

At a presidential address before an international conference of statisticians in Sydney, Berkeley professor Bin Yu proposed in 2014, "Let us own data science."[48] In the early 2010s, journalists, consulting firms, and thinkfluencers were celebrating data scientists at the sexiest job title of the decade. And yet the very academic field closest with understanding data— statistics—seems to have been left in the dust, old-fashioned, perceived as the wrong approach. Statisticians, she said, needed to become more engaged in computing, in contemporary forms of large data, and practices of communications.

In making her case, Yu explained, "Many of our visionary statistics colleagues saw data science coming." Yu was not wrong: a rich tradition of statisticians had focused on data, the potential of computation, and real-world applications. But they had largely swum upstream, against the aggressive mathematization of statistics we encountered, the same anti-empiricist spirit that dominated in symbolic AI. These renegade statisticians tended to have dual citizenship, within the academy and without, usually in industry and government-sponsored research centers.

Alongside John Tukey, no one better embodied renegade statisticians than Leo Breiman, who shuttled from academia to industry. Upon moving from industry and defense work back into academia at UC Berkeley, he was startled. He later described it as being in *Alice in Wonderland*.

> I knew what was going on out in industry and government in terms of uses of statistics, but what was going on in academic research seemed light years away. It was proceeding as though it were some branch of abstract mathematics.[49]

Having left a promising career in mathematical statistics at UCLA, he took on a wide range of statistical work for the Department of Defense and the then-new Environmental Protection Agency. Working outside of academic statistics on subjects such as pollution and tracking Soviet submarines, he explained, he came to focus on prediction over making causal claims using models or doing rigorous hypothesis testing.[50] Outside of academia, Breiman underwent—or perhaps cemented—a fundamental shift in his epistemic values and mathematical practices, away from explanation to prediction.

Statistics was born from making sense of data about diverse populations and systems analyzing data, and yet in the eyes of practitioners like Breiman, the discipline had gone far astray; only then, around 2000, was statistics beginning "to 'recover' from what he called its 'overmathematization' in the post–World War II years."[51] Attendant upon this change in practice, he described a radical contrast between a "data modeling culture" used by an estimated "98% of all statisticians" and an "algorithmic modeling culture," used by "2% of statisticians" but "many in other fields." In the data modeling culture dominating academic statistics, model validation comes through "Yes-no-using goodness-of-fit tests and residual examination." In contrast, the algorithmic culture focused on "predictive accuracy."[52] Restricting oneself to the limited range of models of contemporary statistics was to abandon vast arrays of data, to demand more certain knowledge of causes than often possible, and to limit the creation of new tools needed to solve contemporary problems. Algorithmic culture had too much to offer, even if it meant loosening the traditional demands of statistics.

Breiman was not alone in calling for mathematical statistics to return its focus to real-world data, now with the help of digital computers. In the late 1970s, other statisticians called for their field to more fully embrace the possibilities the digital computer afforded. Despite the growth of computational power, practitioners such as Breiman, Bradley Efron, and William Cleveland argued that academic statisticians failed to face up to large real-world data sets and to integrate computing more centrally within their understanding of the field. In his 1993 call for a "greater statistics" that would learn from data, John Chambers of Bell Labs worried that the overly insular mathematical drive of statistics was "limiting both the influence of statistics and the benefits the field had provided to society."[53] The explosion of data had created the opportunity for statistics to serve an essential function in ensuring rigor and inspiring new methods, but the field was failing to leverage that possibility. Two major computationally oriented statisticians, Walter Stuetzle and David Madigan, called for a dramatic upending of graduate education in statistics, focused on different disciplinary identity.

> Statistics has primarily focused on squeezing
> the maximum amount of information out of limited data. This paradigm is rapidly diminishing in
> importance and statistics education finds itself out
> of step with reality.[54]

Statisticians had much to offer to machine learning and data mining, if they would let themselves.

Statistics departments took note of the rise of data science, as suggested by the statistician Bin Yu in her talk "Let Us Own Data Science." Recognizing the gulf between the world of statistics as taught in universities and the world of data science, she called for far more than an exercise in rebranding. "Data Science represents an inevitable (re)-merging of computational and statistical thinking in the big data era. We [statisticians] have to own data science, because domain problems don't differentiate computation from statistics or vice versa, and data science is the new accepted term to deal with a modern data problem in its entirety."

"The Rise of the Data Scientist"

When Jeff Hammerbacher wrote his description of data scientists in 2009, it combined the mindset of Cleveland's proposal of 2001 and the commercial scale of 1990s data mining with a rapidly emerging toolset for democratizing "big data" in the early 2000s. Mindset and toolset alike were informed by industrial lessons learned, at Facebook in its early days of growth, teams at Bell Labs working since the dawn of digital computation to make sense of the world (and advance the bottom line) through data, and the many efforts to scale data analysis in everyday business work.

The beginning of the millennium also gave rise to a dramatic drop in the cost of computation via cloud-hosted computing, itself facilitated by the information infrastructure of the internet, which allowed data to flow from a computer anywhere in the world back and forth to compute centers thousands of miles away. Echoing the data mining moment of the 1990s, this encouraged companies to turn their web logs, streams of commercial transactions, and customer records into stores of data in the hopes that they could be "mined" for profitable pattern discovery. Protections for consumer data, developed over decades in sectoral regulation applicable to health or financial data, for example, had little authority over the practices of many online companies, even when personal data was being ingested and analyzed. An additional benefit to companies from the cloud was the invisible or "ghost labor," just discussed, where the actual worker could be anywhere on the planet.

Industry-facing publications such as *Harvard Business Review* and O'Reilly Media began to sing the praises of machine learning methods applied to these large troves of transactional data, promising riches and disruption to those who would refine and process the new oil. The enthusiastic adoption of data science in industry was facilitated by the earlier advances in (and marketing of) data mining, big data, and predictive analytics. In short order, a variety of service providers and start-up companies were born to sell digital pickaxes to the new miners, and the gospel of data science filled their marketing materials, turning the flywheel of data exuberance and encouraging companies old and new to reconsider their data strategy and staffing.[55]

The data scientist became a clear job description thanks to start-ups of Silicon Valley, and has now become woven, at times begrudgingly, into the fabric of research and higher education. In some universities, data science is a new institute; in others, it thrives as a renaming of existing departments. Long cold to the manifestos of Tukey, Cleveland, Breiman, and others, some statistics departments began to rename themselves as departments of statistics and data science, as at, for example, Yale and Carnegie Mellon, both in 2017.

Based on their ethnographic observations of practicing data science teams, Gina Neff and coauthors argue, "making sense of data is a collective process."[56] In some ways this has met Chambers's 1993 dream of "greater" statistics, though at the risk of being a field about and of everything, as Jennifer Bryan and Hadley Wickham warned in their "Data Science: A Three Ring Circus or a Big Tent?"[57] As with the terms "artificial intelligence" and "machine learning," the job description "data scientist" proved itself to be a moving target. A Reddit post in the "data science" subreddit asks: "Are Data Scientists at Facebook really Data Analysts?"[58] Along with this drift came a proliferation of job titles as the practice of data became ever more specialized. Today one has not only data analysts and data scientists, but also data engineers, analyst engineers, and, reflecting the growing policy and ethical implications of data, the professional function of "data governance."

Data science has come to encompass a democratization not only of technological tools—high-level statistical software and powerful computing are now easily available—but equally of skills. Just as too many social scientists uncritically used p values from the 1950s forward, researchers in many disciplines have begun using the full range of data science tools, but not always with care. The ease of use of these technologies allows, however, much work that is far from reflexive or critical or transformative.

The challenges of COVID-19 encouraged researchers worldwide to attempt to apply machine learning to predicting the course of the pandemic, with mixed results. In a critical assessment of the field's predictive power, including his own work, Ryan Tibshirani of Carnegie Mellon's Department of Statistics and Data Science concluded, "We as a community missed every surge (meaning, didn't anticipate them)."[59] In a more problematic vein, around 2017 the long-derided "science" of determining human character through the analysis of the face—physiognomy—returned in the form of machine learning studies.[60] Statistics has always had a close and often troubled relationship to the analysis of human difference. Despite devastating criticism of these pseudosciences, researchers since Quetelet and Galton have sought to classify people using statistical means, and they have long sought technical means to distinguish nature from nurture—to find the true geniuses—and the true criminals. Attention-grabbing headlines regularly feature machine learning preprints bordering on pseudoscientific physiognomy. Luke Stark and Jevan Hutson argue, "artificial intelligence and machine learning can now purportedly predict whether you'll commit a crime, whether you're gay, whether you'll be a good employee, whether you're a political liberal or conservative, and whether you're a psychopath, all based on external features like your face, body, gait, and tone of voice."[61] Not just making poor use of machine learning, these works give old-style scientific racism a new objective sheen.

The point is not to condemn data science tools—it's to use them more appropriately, and with an appreciation for their limits. When Mahalanobis improved Pearson's tools in looking at caste in India, he adopted a far more critical approach to the conclusions one could reach. In the early 2010s, both of us had the privilege of teaching an inaugural program in data journalism at Columbia where we sought to teach astonishing young journalists from around the world the critical use of data science technology, to check governments and corporations alike, using careful data analysis and collection, algorithmic analysis, and visualization; we have optimism for the way these tools enable and empower critical investigative work. In their *Data Feminism*, the scholars Catherine D'Ignazio and Lauren Klein illustrate how researchers can make critical use of data science tools to have liberatory potential, rather than rehashing bad old pseudosciences in new scientific garb yet again.[62]

As the scope of data science has expanded, so has the

realization that data can be a powerful force when applied not only to playing games of chess and *go*, or distinguishing photographs of dogs and cats, but when applied to human problems when harms and justice are at risk. Said otherwise, part of the expansion of "making sense of the world through data" under the banner of data science has been the increasing recognition of data's ethical, political, and social impacts.

Ethics without Expertise

"As scientists in the field of data mining," Sergey Brin, then a graduate student, later to co-found Google, wrote a listserv on November 10, 1997, "it is important for us to periodically take a step back from the technology and consider the ethics of using it." He offered a few examples:

> auto insurance companies analyse accident data and set insurance rates of individuals according to age, gender, vehicle type, . . . If they were allowed to by law, they would also use race, religion, handicap, and any other attributes they find are related to accident rate. Health insurance companies also use similar data. . . . All of these can be seen as results of data mining and they have a significant affect [*sic*] on people's lives.[63]

He asked his colleagues to "Please bring your opinions and any relevant examples or studies." Data mining, and later data science, were highly interdisciplinary; they had limits to whose expertise was called upon. While the Stanford meetings were often interdisciplinary, it doesn't appear anyone trained in ethics attended, any more than construction engineers or biologists attended sessions dedicated to their disciplines. One can only wonder if the conclusion of the discussion was "Don't be evil." Ethics, no matter how well considered and well intentioned, tends not to scale well. People in the data mining culture at Stanford knew well how to scale algorithms; they knew how to draw up industry; they knew how to prompt academic research toward practical ends. As for scaling ethics, that was, fair to say, less well covered.

PART III

PART III

CHAPTER 11

The Battle for Data Ethics

Three basic principles, among those generally accepted in our cultural tradition, are particularly relevant to the ethics of research involving human subjects: the principles of respect of persons, beneficence and justice.

—*The Belmont Report*, 1978

What I always see in the AI literature these days is "ethics." I want to strangle ethics.

—Philip G. Alston, John Norton Pomeroy
Professor of Law, NYU, AI Now 2018 Symposium[1]

I n early 2020, Google formed an AI ethics research team, led by two prominent early career scholars in the field. With major academic and popular publications, Dr. Margaret Mitchell and Dr. Timnit Gebru were known for illustrating the potential and real harms of artificial intelligence and suggesting constructive ways to mitigate these harms. Among other groundbreaking work, Gebru had earlier shown, along with Dr. Joy Buolamwini, that several common "commercial gender classification systems" exhibit "substantial disparities in the accuracy of classifying" different demographic groups, particularly "darker-skinned females . . . the most misclassified group."[2] Mitchell was well known for projects on "debiasing" machine learning, as well as collaborations with Gebru including their work on "Model Cards for Model Reporting . . . as a step towards the responsible democratization of machine learning and related artificial intelligence."[3] By summer 2020, the company was ready to offer Google's approach to AI ethics as a service: "Google Offers to Help Others With the Tricky Ethics of AI," according to a headline in *Wired*, "After learning its own ethics lessons the hard way, the tech giant will offer services like spotting racial bias or developing guidelines around AI projects."[4]

Google had succeeded in the face of many previously insurmountable data problems—search, computer vision, even machine translation. Might it soon make progress on this thorny topic? As 2020 came to a close, the vision of an enlightened AI ethical team, harmoniously integrated into the decision-making framework of the corporation, had collapsed. In November of that year, Gebru announced that she had been fired by Google; early in 2021, Mitchell made a similar announcement. Google claimed that Gebru had resigned over a dispute concerning the quality of a research publication; she countered that she was fired for demanding that Google admit to the potential ethical harms of large language models, one of its core technologies.[5] A flurry of public pronouncements by these researchers and their former employer exposed just how vast the gulf between ways AI researchers, even in the same company, might expect a company to integrate ethics in AI research and the realities of product development in a corporation whose profit model rested precisely on the massive use of data on people.

"Evolving the IRB"

This was not the first time a platform company had endured public scrutiny over ethical lapses. An instructive case study arrived in 2014 with the publication of the "emotional contagion" research paper by Facebook researchers.[6] Negative reaction in the press was damning, with headlines such as "Facebook Deliberately Made People Sad. This Ought to Be the Final Straw"[7] and "Users Angered at Facebook Emotion-Manipulation Study."[8] Of more material concern to Facebook leadership were a Federal Trade Commission (FTC) complaint filed by the Electronic Privacy Information Center and a formal request by Senator Mark Warner that the FTC investigate the research. Suddenly, moving fast and breaking things had led to the possibility of profit-curtailing regulation.

Facebook responded with a preemptive, self-regulatory move: explicitly bringing research ethics to Facebook by "evolving" academia's principles-based institutional review board (IRB) process for the corporate environment. Since many researchers within large information platform companies come from academic training, the IRB conception of applied ethics—around comprehensive principles which are interpreted by an adjudicating body—has colored the thinking of many in the technology community. As Matt Salganik, professor of sociology and former member of Princeton's IRB, writes, "the principles-based approach is sufficiently general that it will be helpful no matter where you work (e.g., university, government, NGO, or company)."[9]

In response to public debate about the emotional contagion study, Facebook's Molly Jackman and Lauri Kanerva published in 2016 "Evolving the IRB: Building Robust Review for Industry Research," documenting an approach expanding the applied ethics of institutional review boards. Building on Kanerva's ten years of experience leading the nonmedical IRB at Stanford, the article set out an organizational review board process for Facebook. To be sure, creating an IRB does not cure all ills and prevent all ethical debates. In the emotional contagion study, in fact, a university IRB had deemed the research not necessary to review as it involved no "human subjects," at least none visible in the way understood by research review boards accustomed to medical and social scientific research. So great, however, was the ensuing relitigation of this decision that the editors of *PNAS*, one of the foremost scientific journals in the world, issued a rare "statement of ethical concern" about the paper.[10]

While there have been many attempts to define principles for ethics, none has had near the impact of the human subjects research approach, the domain in which IRBs were first proposed, and itself developed in reaction to scientific scandal. To understand the context computational social scientists have used to frame applied ethics, and to contrast with the diverse set of principles, postures, and products recently offered by dominant technology companies, it's useful to revisit its origin in the 1970s in the form of the *Belmont Report*. This report serves as a foundational document for defining applied ethics, along with the resulting process for ethical review, namely, the creation of institutional review boards (and thus the IRB-inspired organizational review boards such as Facebook's). This institutionalization of ethics has been an influential backdrop for more recent discussions of how to define the ethics of data-empowered algorithms.

From Tuskegee to Belmont

The road to Belmont was paved with the research aspirations of the doctors and scientists of the US Public Health Service; it would end on the front page of *The New York Times*, exposed as an ethical failure so racist and scientifically flawed that it would launch years of work by an interdisciplinary team to craft federal legislative response aimed at ensuring that taxpayer dollars would never again fund such a catastrophe.[11]

On July 26, 1973, the "U.S. Public Health Service Syphilis Study at Tuskegee" made the front page of *The New York Times*, with the headline "Syphilis Victims in U.S. Study Went Untreated for 40 Years." The American public came to know that taxpayer funding had, for decades, supported a study that systematically denied African American men in Tuskegee, Alabama, treatment for syphilis. The experiment was scientifically useless as well as deeply racist. The end of Tuskegee coincided with a time of deep distrust in the US government.

The American Medical Association meeting of May 1936 featured a lecture on untreated syphilis in Black American men. A population of untreated people from the Tuskegee area, in the words of the medical director of the US Public Health Service, "seemed to offer an unusual opportunity to study the untreated syphilitic patient from the beginning of the disease to the death of the infected person."[12] The results were clear: treatment has dramatically positive effects. Taking these experimental "opportunities" meant decades of suffering. The untreated men were kept that way—given placebo treatments, prevented from getting treatment for several more decades, even given draft deferments to keep them from getting treatment during World War II.[13] In the "Tuskegee study," as this effort became known in the medical community, an inexorable scientific logic, prioritizing potential scientific knowledge, clashed with ethical considerations of justice and respect for individual's informed autonomy. The decisions to initiate and continue the experiment reflected as well the long-term grip of structural racism and eugenic thought in the twentieth century. The project continued, publishing its results regularly, until 1972, when a whistleblower, conscious of Nazi and Japanese wartime experimentation, pushed the story into the public eye.

In the wake of its explosive recognition, the US Congress set up a commission "to identify the basic ethical principles that should underlie the conduct of biomedical and behavioral research involving human subjects." The diverse group of commissioners included researchers, lawyers, philosophers, and a former Catholic priest.[14] They were charged with devising an ethical framework for research as well as with designing a process to ensure that this framework would guide and constrain the behavior of researchers. The resulting report established an approach to ethics that combined academic philosophy, social norms, and the realities of the research process. While their motivating problems of the day such as research on children, fetal research, and research on the incarcerated may seem different from those of data-empowered algorithmic decision systems, the commission aimed to provide a framework useful in research more generally. The report entered the *Federal Register* on April 18, 1979, almost six years after the first journalistic exposés on Tuskegee.

The report insisted that research studies involving people could no longer simply be justified by their claimed long-term benefits to society as a whole. Research protocols must carefully weigh the impact of the study on each person involved in it: "the risks and benefits affecting the immediate research subject will normally carry special weight." And the commission warned about taking advantage of oppressed and disempowered groups:

> Certain groups, such as racial minorities, the economically disadvantaged, the very sick, and the institutionalized may continually be sought as research subjects, owing to their ready availability in settings where research is conducted. Given their dependent status and their frequently compromised capacity for free consent, they should be protected

against the danger of being involved in research solely for administrative convenience, or because they are easy to manipulate as a result of their illness or socioeconomic condition.[15]

The report led to limits to research on human beings and a robust set of institutions that implement those limits, however imperfectly. Making ethics stick meant the government first had to sanction some account of ethics, and then design a process, enforced by law and robust bureaucracies, that would guide and constrain ethical research and sanction misuse and abuse.

The resulting framing of ethical experimentation was captured in the *Belmont Report*.[16] Ethics, as defined by the commissioners, enshrined the tension between means against ends (or, as philosophical frameworks, *deontology* against *consequentialism*) and insisted on justice, including the fair allocation of benefits and harms across communities. Rather than setting a specific set of rules or a single maxim, the *Belmont Report* sets out ethics as a negotiated resolution of these tensions, with three *principles* as the shared epistemic backstop—the consensus on which all parties can agree, even when disagreeing about specific applications:

1. *Respect for personhood:* the idea that individuals' autonomy should be respected;
2. *Beneficence:* minimize risk of harm to individuals, maximize public benefit;
3. *Justice:* fair distribution of risk and benefits.

In popular culture, ethics is often conceived of as a philosophical argument or perhaps a small checklist of items, or perhaps even a single maxim. However, the approach of the Belmont group was instead one of "principlism." The idea of principlism is to define a small set of principles with enough generality that they will be applicable not only to the present concerns but likely to future concerns. In the *Belmont Report* itself, the authors explicitly state that they aim their principles to be "comprehensive," meaning that they anticipate their utility for all future applied ethics problems in human subjects research. But every case is different. How can any set of principles apply?

As with a governing document such as the United States Constitution, the value of the documented set of principles is in the community that must strive to interpret these principles as more context-specific standards, and eventually to create unambiguous rules specific to individual cases. Like the Constitution, the *Belmont Report* itself functions as a guide so general that everyone in an organization or a community can agree on its legitimacy. But the power and utility of the document is limited by the existence of a community which does the hard work to distill these principles into standards, rules, and therefore into practice.

Principlism is not intended as an algorithm or checklist, yielding a clear or automatable decision. Instead, the principles are *meant* to be in tension, a productive tension that provides a common vocabulary and rubric for adjudicating difficult decisions. This common language and common set of values serves a powerful social function: ensuring that members of a community, such as the employees of a company or the users of a product, feel that the decision was at least made legitimately and with a healthy process, even if the result is not one with which everyone will agree.

Major Principles in Belmont

While the principles identified by the Belmont commission drew on centuries of ethical philosophy, the commissioners took the principles, with all their tensions, to be present in existing social norms. In their view, "the national commission almost certainly believed that these principles are already embedded in preexisting public morality."[17]

A central concern of the Belmont commissioners was how to balance the collective good that might come from a scientific experiment with the impact on each of the individual research subjects. The commission's report was designed to capture the tensions between legitimate ends and means, enshrined in the two first principles of "respect for persons" and "beneficence."

Respect for persons requires respecting the autonomy and dignity of individuals participating as research subjects. Often instrumentalized as "informed consent," the principle derives from the deontological tradition within the philosophical ethics, strongly associated with Immanuel Kant. In the context of human subjects research this demands ensuring the informed consent of those with diminished autonomy, such as children or the incarcerated.

Beneficence includes to weighing the potential benefits and harms of a research project. Often this is summarized as "Do no harm," but more generally this refers to maximizing benefit and minimizing harm not only to research subjects but to society. More recently this principle has extended to harms beyond human society, for example, to other living creatures or to the environment. This principle itself derives from the consequentialist or utilitarian philosophical tradition, associated with John Stuart Mill, Jeremy Bentham, and others.

This principle is particularly challenged by algorithmic ethics, in that complex algorithms make speculating on the possible unintended effects and potential harms difficult. On the other hand, algorithmic products and services like recommendation engines also make possible the monitoring and mitigation of such harms as they are revealed. Unlike a defective product which must be recalled and repaired, an algorithm can be tuned and digitally redeployed.

The third principle of Belmont is justice, focused not on the tension between ends and means but on norms of fairness. Particularly in the context of research on the incarcerated, commissioners were concerned not only with equal treatment but oppression and maldistribution. Reflecting upon her role on the commission some years later, in 2004, Professor Karen Lebacqz underscored the commission's commitment to justice. She explained this commitment could now be cast in a stronger idiom. "We talked about justice and we talked about it primarily in the language of equal treatment and protection of the vulnerable. A language that we did not use in those days but that has become very prominent since and very important to me, is the language of oppression." She underscored this way of speaking would bring out more clearly what a commitment to justice in research would entail. "I think there is a difference between populations who are simply vulnerable and populations who are oppressed. And, justice requires rectification of oppression and that might set some structures differently than the way that we did so many years ago."[18]

The three general principles are taken to imply additional ethical standards. Privacy, for example, can be viewed as an example of informed consent—where privacy is understood as circumstances around a disclosure of the fact, rather than the fact itself. For example, we may consent to share a fact with a doctor that we would not share with our teachers or students. Similarly, "fairness" is viewed as a fundamental aspect of justice. Fairness aims to avoid, for example, medical experiments on the poor and disenfranchised, who suffer the risks of these experiments, whereas the benefits flow toward the empowered who can afford the resulting medicines or medical treatments.

These three principles were designed to be "comprehensive" for covering applied ethical problems of human subjects research, but applying ethics means a shift in power to enforce. Alongside their philosophical work, the commissioners proposed a codification of the Institutional Review Board process in law—a potent institutionalization of ethics within organizations.

Institutionalizing Principlism from IRB to Silicon Valley

Along with the foundational principles, the commission published—separately—a 132-page proposal for the creation of process design to operationalize them.[19] These guidelines shaped the creation of the institutional review boards that would govern human subjects research at all US universities as a condition of federal funding. Federal funds may be directed to a project only after passing review by such a board, whose deliberations are guided by these shared principles. Controlling funds gives IRBs rules power, power that mere regulations might never have. However imperfectly— and there is no question the story of IRBs is replete with imperfections—these boards are designed to ensure that researchers act in accord with the principles. In the intervening years, IRBs have had to apply these principles to new technical topics, including genetic engineering and more recently research in computational social science.

The IRB model continues to serve as a major model for institutionalizing ethics. While we may disagree with the ethical framework, the rules, or the institutionalization of the contemporary IRB system, the key point is that the system brings together a rich philosophical account of ethics, a means of putting that philosophical reflection into practice in nuanced cases, and institutional means for enforcing the framework. Ethics by itself is toothless; regulations without ethics are mere bureaucracy.

Design is "the intentional solution to a problem within a set of constraints."[20] The IRB is an example of process design. All examples of design, of course, are also statements of power: whose intent is respected, who solves the problem, and who sets the constraints. The resulting design, whether a product or a process, also impacts power by rearranging who can do what to whom.

As social scientists, activists, computer scientists, and journalists increasingly began to signal the potential and realized dangers of large-scale automated decision systems—machine learning in practice—to communities and to democracy, the Belmont principles and the IRB structure offered a powerful existing system for anyone looking to gauge the impact of technologies and organize them toward ends other than, say, profit. For companies like Facebook, they offered a framework for self-regulation, rather than government regulation. As the case of Facebook's "evolving the IRB" suggests, the first challenge for companies aiming to adopt these principles is one of recontextualizing: How do these principles, developed in the case of human subjects research, apply to information platform companies? The second challenge for companies aiming to adopt these principles is one of design and the distribution of power around decision-making: How do companies institutionalize organizational design and process design such that these principles meaningfully constrain and guide decisions?

Evolving the IRB has proven not to be straightforward or effective. In particular, the centralization of ethical concerns into one group who "owns" ethics at Facebook has not stemmed the growing chorus in the years since calling for data ethics governing such companies, where data-empowered algorithms have the greatest impact. The framing of ethics over the past decade in terms of "principles" has not meant that data-powered platform companies pivoted, organizing their businesses by hiring philosophers. Nor is it clear that doing so would have any practical effect. Principlism presumes that the principles may be in tension with each other, requiring good faith adjudication by individuals who share the principles as their common vocabulary and values. While common ethical principles are grounded in a philosophical tradition, Jacob Metcalf, Emanuel Moss, and danah boyd conclude from an ethnographic study of employees at platform companies that ethics should be better understood "as social phenomena and not as primarily philosophical abstractions."[21] The realization of ethical process requires broad buy-in by the members of the organization and the empowerment to make organizations comport with principles applied to their business practices. Unmoored to the structure of the IRB and its central role as a "pause point" in controlling research funding, it is unclear how to convince colleagues to value ethical principles and how to design organizations and processes within which the principles could serve as any constraint—particularly if those constraints would reduce profit.

In at least one respect, Facebook's approach turned out to be a great success. Surveying the state of AI ethics in 2019, Oxford researcher Brent Mittelstadt found a "convergence of AI Ethics around principles of medical ethics . . . historically the most prominent and well-studied approach to applied ethics." It was also a successful approach for Facebook in that it "provide[d] policy-makers with a reason not to pursue new regulation."[22]

As the challenges of "unethical" behavior came to draw attention from social scientists and, increasingly, the press, scholars questioned whether ethics, often in bedrock laid by the Belmont principles, and their ensuing institutional design for governing human subjects research could provide any shield against the rising harms and injustices of corporate uses of personal data.

Owning Ethics: Process, Organization, and Power

Moving beyond calls for ethical principles, the past few years have seen a widening call for spelling out what a meaningful ethical *process* within a corporation might entail. As one example, Inioluwa Deborah Raji and coauthors, including Margaret Mitchell and Timnit Gebru, with whom we began this chapter, argue for a process for algorithmic auditing.[23] Earlier, philosopher Shannon Vallor, then of the Markkula Center of Applied Ethics, building on the business ethics literature, developed a set of checkpoints in the development of a digital product, in which different questions are asked at different points in the product development.[24] These processes—audits and checkpoints—couple ethical interrogation to moments of decision beyond which the scale of impact (including harms) increases. These decision moments can also be framed as moments where power is exercised—where ethics is given "teeth." And they go beyond considering the firm as a uniform whole. Without strong coupling between ethical decision making and the objectives of individuals in corporations—such as promotions or successful product launches—it is unclear how ethics becomes integral to the community.

Complex companies necessarily comprise separate teams with separate interests. These interests may include "ownership," meaning responsibility for a particular revenue stream or a user behavior. It is often unclear where in an organization to locate a team responsible for, in corporate parlance, "owning" ethics.[25] Ensuring individuals will integrate audits, pause points, or other ethical checks into practice requires an alignment between shared principles and the incentives of diverse actors within a firm. To quote an anonymous tech employee, "the system that you create has to be something that people feel adds value and is not a massive roadblock that adds no value, because if it is a roadblock that has no value, people literally won't do it, because they don't have to."[26] This successful alignment is what business ethics scholars Theodore Purcell and James Weber described in 1979 as "[t]he institutionalization of ethics . . . getting ethics formally and explicitly into daily business life . . . into company policy formation . . . into all daily decision making and work practices down the line, at all levels of employment."[27]

The case of Google underscores the gap between aspiration and implementation. Even before the high-visibility reorganization at Google's Ethical AI team, Google had earlier stumbled at positioning itself as an ethical company. For example, the company had created an external council to advise on ethical implications of AI in March 2019, then hastily disbanded it the next month after internal and public criticism about its makeup as well as its integration into the company's decision-making processes. Former Google employee Meredith Whittaker dismissed these moves as "ethics theater" in 2018, asking "Can they cancel a product decision? Do they have veto power otherwise?"[28]

Google is not alone. The lawyer Ben Wagner accused tech companies of "ethics washing"—working to avoid regulation without meaningfully defining ethics and designing processes to guide decisions using ethics. In his "Ethics as an Escape from Regulation: From Ethics-Washing to Ethics-Shopping?," Wagner proposes six criteria for ethical processes in companies:

1. External participation of relevant stakeholders.
2. External and independent oversight.
3. Transparent decision-making process.
4. A stable list of standards, values, and rights.
5. Ensure that ethics do not substitute [for] fundamental rights or human rights.
6. "A clear statement on the relationship between the commitments made and existing legal or regulatory frameworks, in particular on what happens when the two are in conflict."[29]

The final criterion contrasts ethics with law: while legal tradition stretches for thousands of years and shapes the process and legitimacy of government, applied ethics aims to form consensus among stakeholders as to the legitimacy of decisions, particularly those made by those in power. Increasingly, power is in the hands of data-empowered, internationally active technology companies; the stakeholders include citizens across the world, as well as leaders of state searching to craft regulation to respond to this shift in power.

As we will explore in our final chapter, individual employees advocating for ethics have many "people power" tools at their disposal. Often, they find that ethical practices sharply conflict with the financial goals of their employers. Nonetheless, corporations are increasingly interested in the appearance of ethics. As Metcalf and colleagues write, "Ethics is arguably the hottest product in Silicon Valley's hype cycle today."[30] Much of this is in reaction to threats of increased regulation and "state power," as well as "people power" such as internal critiques by their employees.

The Limits of "Tech Fixes"

What if ethical issues could be addressed through techno-logical fixes, rather than complex social, deliberative action? Many in the technical community have sought just such solutions. Within the community of technologists hoping to advance algorithmic ethics, two particular facets—fairness and privacy—have blossomed into areas of technical research in the past decades.

Most MIT graduate students in computer science don't publish in *The Journal of Law, Medicine & Ethics*. Latanya Sweeney was clearly not one to be constrained by conven-tion. Concerned about the illusion of anonymity provided by merely removing the "name" field from publicly released databases, she convincingly illustrated in a series of papers how a database considered "anonymous," thanks to having names redacted, could be combined with a second database with other unique identifiers to reidentify individuals and thus expose sensitive information. Peer-reviewed papers may be coin of the realm in academia, but don't nearly have the impact of a real-world experiment: for this, Sweeney illustrated the point by reidentifying her state's governor using his own "anonymous" medical record, revealed by combining with public voting records.[31] The identifiers common to both (birthdate, gender, and zip code) together formed the key to this lock. A few years later, Arvind Narayanan and Vitaly Shmatikov similarly showed how to deanonymize at least some of the reviewers in the Net-flix Prize data set using data from another database, caus-ing Netflix to pull the data set.[32] Such deanonymization threatened to reveal highly personal preferences that could embarrass and even endanger users.

Sweeney proposed a technical defense against such an attack: k-anonymity, an attribute of a database in which no record is unique, but is identical to at least k-1 other such records.[33] For example, in the voting record we could have released only birth month (rather than date), or only the first three or four digits of a zip code (rather than all five), until we ensure that any one record cannot be uniquely identified. Intuitively, such a process provides a level of plausible deniability: "It wasn't me; it was one of the k-1 other identical records in this database!"

A similar technically plausible form of deniability is that of *differential privacy,* an aleatory approach to providing privacy. Motivated in part by "the spectacular privacy compromises achieved by Sweeney," Cynthia Dwork proposed differential privacy in 2006 as a noise-generating technique, such that the original database is never revealed.[34] As with k-anonymity, in which we must choose the granularity k desired, differential privacy comes with immediate subjective design choices as to the strength of the noise to be injected, as well as the noise model itself (e.g., if we are to query a database for words contained in a document rather than heights of patients, we would choose a different noise-injecting mathematical model). Such a choice of granularity illustrates the tension between privacy and utility. As Dwork wrote in the original work proposing the technique, privacy "requires some notion of utility—after all, a mechanism that always outputs the empty string, or a purely random string, clearly preserves privacy."[35] Differential privacy has continued to be refined, developed, and extended, with a flurry of attention over the past years thanks to the decision by the US Census Bureau to use differential privacy when releasing records from the 2020 census.

In our first chapter we met the computer science researcher Hanna Wallach via her 2014 talk at the workshop named "Fairness, Accountability, and Transparency in Machine Learning" (originally FAT-ML, now FAccT). Over the course of the next few years, as this community grew, more technically oriented practitioners from computer science focused on developing highly mathematical and technical approaches on defining and quantifying *fairness*—more code, less philosophy and law. In parallel to the blossoming of this technical literature, a literature on the dangers of algorithms grew, including germinal works by Cathy O'Neil, Virginia Eubanks, and Ruha Benjamin, among others.[36] While notions of fairness have been part of US law for decades (particularly after the Civil Rights Act of 1964), fairness had figured only slightly in the technical literature until recently. As this literature rapidly expanded, some particular surprises shaped the goal of applying engineering mindset to problems of fairness. The first was that there are many plausible quantitative definitions of fairness; the second was that some of these definitions are mutually incompatible—both formally and in practice.

To illustrate the challenge of quantifying fairness, consider the case of "Machine Bias" from May 2016.[37] The piece by the journalism nonprofit ProPublica investigated COMPAS, a proprietary algorithm developed by the company Northpointe to predict criminal recidivism in Broward County, Florida. Careful work showed that the algorithm was unfair in the sense that the group of white defendants who were algorithmically scored to be "high crime" were in fact more likely to go on to commit crimes than the similarly scored Black defendants. However, three researchers at Northpointe published their own analysis just two months later, showing that its methods were fair in the sense that the algorithm is "equally accurate for blacks and whites." Princeton computer scientist Arvind Narayanan has illustrated how different technical definitions have radically different politics, using twenty-one alternate accounts of fairness.[38] Along with Narayanan, Solon Barocas and Moritz Hardt, two of the co-organizers of the original FATML workshop, summarize three central fairness measures, as they might be applied to questions of racial discriminations:

> Independence: the model output is independent of (in the nineteenth-century language of Galton, "uncorrelated with") race

> Separation: given the true outcome (for example, when defendants who in fact did or did not commit later crimes are considered as separate groups), the algorithm's score is independent of their race

> Sufficiency: given the algorithm's score (for example, when defendants who were predicted to commit or not to commit later crimes are considered as separate groups), the true outcome is independent of race.

These conditions can be stated more mathematically, and for general cases of protected attributes (i.e., not necessarily race) and general outcomes.*

The ambiguity over *which* definition of fairness to use complicates what would otherwise be a well-trodden path for machine learning: statistical optimization. As we saw in our chapter on machine learning, computational methods in machine learning are by now well proven at learning what policies can optimize a desired objective, even in complex environments combining algorithms, digital products, and societies. These methods can work even when the objective includes a competition, for example, between a

---

* For these definitions in more mathematical terms, see Solon Barocas, Arvind Narayanan, and Moritz Hardt, "Fairness and Machine Learning," 2019, https://fairmlbook.org/.

"metric" (e.g., statistical accuracy) and a "countermetric" (e.g., model complexity).[39] The computer science researchers Michael Kearns and Aaron Roth advocate this approach in their recent book *The Ethical Algorithm*. "The only sensible response to this fact—from a scientific, regulatory, legal, or moral perspective—is to acknowledge it and to try to directly measure and manage the trade-offs between accuracy and fairness."[40] Tech fixes, though, only go so far. For example, even an optimal algorithm—optimized for fairness and accuracy—will not fix problems such as statistically self-reinforcing over-policing, in which the prediction of "crime" and "arrests" are conflated, sending more police to an area in which more arrests have been previously observed.[41] As Kearns and Roth observe, the algorithm is only one part of a socio-technical system: "Good algorithm design can specify a menu of solutions, but people still have to pick one of them."[42]

And even when these technical approaches work—and they often do—they necessarily require power in (or over) organization, with authority to enforce and direct, not merely to critique. As the collapse of the Ethical AI team at Google illustrates, it is unclear where, within the organizational chart, one could place the agency to make such decisions.

For all their importance, such technical solutions focus on reworking facets of algorithmic systems and the collection of data to minimize bias and the effects of structural inequality, but do not function to alter social structures driving and maintaining the inequality. They strive for fairness, rather a more robust pursuit of justice. "A broader focus on *data justice*," write Catherine D'Ignazio and Lauren Klein, "rather than *data ethics* alone, can help to ensure that past inequalities are not distilled into black boxed algorithms."[43]

And given the growing centrality of automated algorithmic systems, justice in our societies itself depends increasingly on data justice. In their critique of technical approaches to ethics, Safiya Noble and Matthew Le Bui likewise argue, "Simply striving for fairness in the face of these systems of power does little to address the ways that digital technologies are increasingly central to other forms of structural power."[44] In crafting AI ethics, too many researchers turned to the most procedural facet of the Belmont report, while missing its concerns with substantive justice in the face of socioeconomic, sexual, and racial disparities.

> The endgame is always to "fix" A.I. systems, never to use a different system or no system at all.
> —*Julia Powles and Helen Nissenbaum*[45]

The aspiration to apply a technical fix to problems in AI presumes that the use of AI is there to be improved, rather than pushed back or even resisted entirely. Lawyer and technology scholar Frank Pasquale identifies the movement to question even the building of systems as a "second wave" of algorithmic accountability: "While the first wave of algorithmic accountability focuses on improving existing systems, a second wave of research has asked whether they should be used at all—and, if so, who gets to govern them."[46] A growing number of technologists are joining lawyers, sociologists, and activists in posing these more structural questions and taking action to enact a reordering of power as private individuals often working together to press corporations and governments toward greater fairness and justice.

(Self-)Regulatory Capture

The private ordering of individuals is increasingly resonant with a growing number of institutions which seek to address ethical concerns in algorithms, either via fixes or by challenging the use of AI. Many of these "self-regulatory" organizations are however themselves funded by the companies they seek to critique, leading to a conflict which can slow, stifle, and subtly direct such criticism. In "The Invention of 'Ethical AI': How Big Tech Manipulates Academia to Avoid Regulation," Rodrigo Ochigame, now a professor at Leiden University but then a PhD candidate at the Massachusetts Institute of Technology, traced the flow of funding from technology companies to research institutions aiming to create a field of "Ethical AI" that would critique and constrain these companies' profit-generating products and services.[47] Both AI researchers and many (perhaps most) ethical AI researchers find themselves profoundly dependent on a small array of corporations. "Enticing researchers," several Australian scholars have recently argued, "to become suppliers of virtue that can be easily dismissed or incorporated . . . offers little resistance to existing forms of business organisation or business models."[48] And yet, as we have suggested before, ethics without power may be inert, and power without ethics lacks any positive social and political direction.

The opacity of algorithmic products and their harms and impacts, as well as long-standing organizational complexities at firms, makes "doing ethics" difficult. The challenges are only amplified by the tension between difficult to quantify long-term ethical concerns and short-term quantitative concerns—typically expressed as an organizing principle around optimization of so-called "metrics." For example, commitments to privacy can be challenged by the profitability of "surveillance capitalism," to use a phrase championed by Shoshana Zuboff: the enhanced tracking of individuals and the economic demand for such granular data for use in marketing and beyond.[49] As a technology, the algorithms driven by such data constitute what Zeynep Tufekci terms "persuasion architectures"—used equally effectively whether the persuasion is in support of a product or a political candidate. To understand the power and profitability of these architectures, we must turn from the view of humans as ethical deciders to humans as sources of valuable attention. In this arena, as the artists Carlota Fay Schoolman and Richard Serra wrote in 1973, "you are the product."

tures have become so optimized in statistical performance as well as ease of use that anybody can do it. As one example, Facebook's "lookalike audiences" allows marketers to ask Facebook to find other people who look like the people who performed a certain action. Marketers can thus target content to new users who resemble demographically or behaviorally prior users who performed an action, like clicking on a particularly provocative link. This can be done without any particular savvy or intuition about market research or user psychology; just a small amount of money for the individual purchaser, though vast in aggregate for Facebook or Google.

Bernays realized, even in the 1920s, that the laws of persuasion apply equally in marketing or in politics. Before Bernays was honing the message for tobacco, he was shaping public perception in politics. In 1924 he arranged for a group of popular celebrities to appear with the "practically inarticulate" president Calvin Coolidge to improve his image. Today, aspiring politicians benefit from the same statistical methods as marketers, leveraging digital tools for market surveys to craft campaigns and messages as well as to target the optimal message to the optimal user. Bernays saw no line between advertising and politics; indeed, he saw this engineering of consent as a good thing for democracy. "Bernays saw this as an unavoidable part of any democracy," according to Zeynep Tufekci. "He believed, like Dewey, Plato and Lippmann had, that the powerful had a structural advantage over the masses. . . . He urged well–meaning, technologically and empirically enabled politicians to become 'philosopher–kings' through techniques of manipulation and consent engineering."[34]

While these techniques can—and must be—used for good, Bernays was clear, they "can be subverted; demagogues can utilize the techniques for antidemocratic purposes with as much success as those who employ them for socially desirable ends." A leader seeking good objectives, Bernays argued, must "apply his energies to mastering the operational know-how of consent engineering, and to out–maneuvering his opponents in the public interest."[35] Even as Bernays's bracingly honest use of the term "propaganda" fell out of favor during the Cold War, the effectiveness of "engineering consent" grew dramatically with the increasing use of data and algorithms to inform and optimize persuasive messaging. It was not long before the techniques perfected for digital advertising online came to inform political messaging.

In 2007, Rayid Ghani, then an employee at Accenture Technology Labs, described an "Individualized Promotion Planning system." He celebrated how data would allow radically new forms of individualized targeting. "In addition to using newspapers, in-store displays, and end caps to highlight their products and run promotions, retailers can influence individuals in a vastly different way using individual consumer models."[36] Business goals can be met by understanding customers at a granular level. This technology allows every company to target every customer as a separate person, not just a representative of a statistical category. Just such targeting is at the heart of marketing and political campaigns.[37] Ghani served as Obama's chief scientist in the 2008 election, in part using these data-empowered views of the electorate as customer segments.

In an op-ed in *The New York Times* soon after the reelection of Obama in 2012, Ethan Roeder concluded with a celebration of the individual at the heart of the Obama campaign's data strategy. "Campaigns are . . . moving toward treating each voter as a separate person."[38] This rosy picture of persuasion is interpreted more darkly after the concerns of the past few years in which this technology has been characterized as too persuasive. Unbounded granularity, deep contextual information about individuals, and personalized persuasion architectures which optimized for engagement had created, many feared, too great a success for democracy to function healthily.

This vision of an engineered public fills some with hope, and others with concern that informed consent has been infringed by those in power. Such tools could easily be put to ill use, and we now know they have. Tufekci warned in 2018, "To microtarget individuals with ads, today's platforms massively surveil their users; then they use engagement-juicing algorithms to keep people onsite as long as possible. By now it's clear that this system lends itself to authoritarian, manipulative, and discriminatory uses," of which she gave numerous examples.[39] But you can't microtarget somebody unless you differentiate them based on who they are or what they've done. And that requires you to have copious data and machine learning. While many feel that our free-market norms are not threatened by powerful companies delivering deeply personalized advertising, our concerns over power are heightened when these abilities empower the state. The algorithm, though, works for states and corporations alike.[40] Matthew Salganik warns us, "These capabilities are changing faster than our norms, rules, and laws."[41] We should add: they are likely changing faster than our analytical tools for understanding the relationship between our social and economic realities and the conceptual worlds, mediated in platforms, through which we experience and act in them.

Kenya's 2017 election, Brexit, and the 2016 US election greatly amplified—and popularized—concerns about algorithmic manipulation. A flashpoint of concern was the firm Cambridge Analytica; then-CEO Alexander Nix shared a worldview in 2017 that echoed and updated that of Bernays:

There's no question that the marketing and advertising world is ahead of the political marketing and political communications world. And there are some things that I would definitely [say] I'm very proud of that we're doing which are innovative. And there are some things which is best practice digital advertising, best practice communications which we're taking from the commercial world and are bringing into politics.[42]

Researchers have reached no consensus on the ultimate effects of this attempted manipulation.

Worrying about the effects of the adtech and persuasion architectures does not require us to believe the claims of advertisers and technical snake-oil salespeople about the effectiveness of their ads. Tim Hwang and Cory Doctorow have brilliantly stressed the deep limits and deceptions around targeted advertising.* While adtech, either for commerce or politics, surely doesn't work in the ways those hawking it suggest, it has dramatically transformed our media landscape and consolidated a landscape of digital advertisers into a near duopoly (Facebook and Google), with unpredictable effects. Facebook and Google don't need the ads to work as promised—they need advertisers to believe they work. Perhaps it's a shell game, but it's one that dominates our informational landscape ever more, for better and worse.

---

* Cory Doctorow writes, "Surveillance capitalists are like stage mentalists who claim that their extraordinary insights into human behavior let them guess the word that you wrote down and folded up in your pocket but who really use shills, hidden cameras, sleight of hand, and brute-force memorization to amaze you." "How to Destroy 'Surveillance Capitalism,' " *OneZero* (blog), August 26, 2020, https://onezero .medium.com/how-to-destroy-surveillance-capitalism-8135e6744d59; see Tim Hwang, *Subprime Attention Crisis: Advertising and the Time Bomb at the Heart of the Internet* (Farrar, Straus & Giroux, 2020).

use them to defend and extend their own power and control. So it takes a while to reorient norms, laws, architecture and markets in a way that harnesses these emerging capabilities in order to empower the defenseless—but it can be done.[41] Technology means change, but societal change takes time: as we've seen, sometimes it takes decades for a technology to get integrated into society before it comports with our values and norms—if it does at all. Many potential forces, large and small, are available to us, directly and indirectly, to shape the relationships among technology and norms, laws and markets, and data's role in it all.

ACKNOWLEDGMENTS

This book emerged from a class we developed for Columbia and Barnard undergraduates. We first envisioned teaching it at the instigation of a group of Columbia students, and with the financial, moral, and bureaucratic support provided by the Collaboratory at Columbia, led by Richard Witten and Jeanette Wing, then director of Columbia's Data Science Institute. Were it not for our students from 2017 to 2022, this work would have been far more narrow and limited in scope. They consistently pushed us to sharpen our account of the persistent conflicts among data, truth, and power, from the eighteenth century to present day. We're grateful for their focus, engagement, and curiosity. Their questions shaped the material and inspired us to work harder to find the historical and technological roots that help "explain" the present yet make the present strange by picturing the counterfactual realities that easily could have been. These other worlds, those that could have been, also inspired our discussions of the possible futures we all may yet create and enjoy.

MY (WIGGINS'S) UNDERSTANDING of the realities of developing and deploying machine learning in industry has been shaped by my fantastic colleagues, past and present, on the Data Science team at *The New York Times*, along with similarly skilled technologists at various information platform companies who would probably prefer I not name them but know who they are.

My understanding of ethics was initially forged from discussions with Matt Salganik about his book *Bit by Bit*, and refined by many discussions with Robin Berjon, director of data governance at *The Times*, particularly during Professor Salganik's year as scholar in residence at *The Times*. I thank David Blei, David Donoho, Gerd Gigerenzer, Mark Hansen, Gina Neff, Peter Norvig, Cathy O'Neil, Deb Raji, Ben Recht, Alfred Spector, Latanya Sweeney, Anne Washington, Hadley Wickham, and Jeanette Wing for additional insights and driving questions on data, data science, and ethics. Understanding the ways algorithms shape and distort reality was helped by conversations for years with David Carroll, Renee DiResta, Joan Donovan, and Justin Hendrix. The broader impact of data on our collective truthmaking, particularly in the arena of media and politics, was greatly clarified by many conversations with Mark Thompson.

I thank Matt Jones, Ariel Kaminer, Rob Phillips, and Allison Schrager for boundless advice on how actually to write a book. And thank you of course and always to my parents, Richard and Carolyn Wiggins.

WE'VE TRIED THROUGHOUT to express our debts to scholars, policymakers, and technologists who have transformed our thinking. I (Jones) have been exceptionally lucky to work with Stephanie Dick, Richard Staley, Mustafa Ali, Jonnie Penn, and Sarah Dillon on the two-year project History of Artificial Intelligence: A Genealogy of Power, a Mellon-Sawyer seminar, and before that with Dick, Penn, and Aaron Mendon-Plasek on a workshop on the history of AI. Throughout the process, inspiring scholars from around the world challenged and inspired us. I benefited greatly from a fellowship at Data & Society, where I had the great fortune to work especially with danah boyd, Darakhshan Mir, Jeanna Matthews, Seth Young, and Claudia Haupt. In addition to great fellowship, Seth Young provided much guidance in thinking about accountability—an unfinished project with him made one of the chapters here possible. At Columbia, Eben Moglen kindly allowed me to take two law courses, and Rachel Schutt and Cathy O'Neil permitted me to attend their first data science course. Chris Wiggins attended my first preliminary talk on these subjects—the clarity of thought and generosity I first encountered there has run through our entire collaboration, culminating in this book.

Key facets of the history of the data sciences presented here came together for workshops at the Max Planck Institute for History of Science in Berlin and at the Huntington hosted by UCLA. More academic iterations of ideas here have benefited enormously from questions and comments from sage audiences at University of Michigan–Ann Arbor; Indiana University; UC Berkeley, UCLA, UCSB, UCSD; Nanyang Technological University, Sciences Po; Cornell Tech; University of Cambridge; the European University Institute; University of Chicago; University of Siegen; University of Pennsylvania; Rutgers; and the Department of Computer Science at Columbia. I've greatly benefited from communities around David Isenberg. My graduate student Aaron Mendon-Plasek has been working deep in the history of machine learning; his work will soon eclipse the accounts given here.

Undertaking this project meant returning to school, something made possible by Mellon and Guggenheim foundations. The Sloan Foundation funded a series of workshops at Columbia on histories of data and of artificial intelligence.

Archivists at Stanford Special Collections, the British Library, the American Philosophical Society, Columbia Special Collections, UMass-Amherst, the Baker Library at Princeton, and the University of Nevada, Reno Special Collections, made the research possible. The many government employees who answer FOIA requests likewise ensured that much of the history here could be told.

With their intense love of books, my three daughters may read this one someday—and, without doubt, uncover the remaining errors and infelicities. Elizabeth Lee's insight, love, and intelligence echo through everything I write and do. 我愛妳們!

WE BOTH THANK Ella Coon and Susannah Glickman, who went far beyond the call of duty in correcting, editing, and improving the final draft of the book. Generous colleagues Stephanie Dick, danah boyd, Theodore Porter, David Sepkoski, and Sarah Igo carefully read and greatly improved draft chapters, noting our errors and encouraging the work. Excellent critical comments on drafts of this book were likewise provided by Chris Eoyang, Su Hang, Willian Janeway, DJ Patil, and JB Rubinovitz. All remaining errors are entirely our fault. The Sloan Foundation supported the completion of this book; thanks especially to Josh Greenberg for his encouragement.

Cathy O'Neil and Mark Hansen invited us in 2014 to teach data journalists as part of Columbia Journalism School's Lede Program, and thus to put into practice critical data science. This was our first chance to develop a curriculum together and to teach together; both the structure—mixing lecture and functional engagement with material in Python—and the questions asked about the role of data in society and in truthmaking informed our later course.

Our agent Eric Lupfer shaped our initial centrifugal ideas into a proposal and then a book with a strong narrative and argumentative focus. With his insight and precision, our editor John Glusman helped turn a first draft into the book before you.

NOTES

PROLOGUE

1. Kevin Roose and Cecilia Kang, "Mark Zuckerberg Testifies on Facebook Before Skeptical Lawmakers," *New York Times*, April 11, 2018, sec. US, https://www.nytimes.com/2018/04/10/us/politics/zuckerberg-facebook -senate-hearing.html.
2. The weekly cadence of the class involved two separate meetings, with discussions on Tuesdays and functional engagement with the material on Thursdays: that is, computational engagement in Python performing the types of data analyses and machine learning models discussed on Tuesdays. We've not attempted in this book to capture the applied half of the class, but we invite readers who are interested in a closer read of the material to engage directly with the data and the code provided online at the course website: https://data-ppf.github.io/.
3. Our language here borrows from Phillip Rogaway, "The Moral Character of Cryptographic Work" (2015), 1, https://web.cs.ucdavis.edu/~rogaway/papers /moral-fn.pdf.

CHAPTER 1: THE STAKES

1. Hanna Wallach, "Big Data, Machine Learning, and the Social Sciences," Medium, December 23, 2014, https://medium.com/@hannawallach/big -data-machine-learning-and-the-social-sciences-927a8e20460d.
2. Wallach.
3. danah boyd and Kate Crawford, "Critical Questions for Big Data," *Information, Communication & Society* 15, no. 5 (June 1, 2012): 663, https://doi .org/10.1080/1369118X.2012.678878.
4. Such advocacy echoes movements in the 1960s among engineers seeking more social and environmentally conscious technologies, chronicled in Matthew H. Wisnioski, *Engineers for Change: Competing Visions of Technology in 1960s America* (Cambridge, MA: MIT Press, 2012).
5. Safiya Umoja Noble, "Google Search: Hyper-Visibility as a Means of Rendering Black Women and Girls Invisible," *InVisible Culture*, no. 19 (October 29, 2013), http://ivc.lib.rochester.edu/google-search-hyper-visibility-as-a -means-of-rendering-black-women-and-girls-invisible/. She developed the arguments in her *Algorithms of Oppression: How Search Engines Reinforce Racism* (New York: New York University Press, 2018).
6. Cathy O'Neil, *Weapons of Math Destruction: How Big Data Increases Inequality and Threatens Democracy* (New York: Crown, 2016), 48.

7. Ruha Benjamin, *Race after Technology: Abolitionist Tools for the New Jim Code* (Cambridge, UK; Medford, MA: Polity Press, 2019), 44–45.
8. Meredith Whittaker, "The Steep Cost of Capture," *Interactions* 28, no. 6 (November 2021): 50–55, https://doi.org/10.1145/3488666.
9. Virginia Eubanks, "Public Thinker: Virginia Eubanks on Digital Surveillance and People Power," interview by Jenn Stroud Rossman, *Public Books* (online), July 9, 2020, https://www.publicbooks.org/public-thinker -virginia-eubanks-on-digital-surveillance-and-people-power/.
10. Lisa Nakamura, *The Internet Is a Trash Fire. Here's How to Fix It*, 2019, https://www.ted.com/talks/lisa_nakamura_the_internet_is_a_trash_fire _here_s_how_to_fix_it.
11. Zeynep Tufekci, "Engineering the Public: Big Data, Surveillance and Computational Politics," *First Monday*, July 2, 2014, https://doi.org/10.5210/fm .v19i7.4901.
12. Renee DiResta, "Mediating Consent," *ribbonfarm* (blog), December 17, 2019, https://www.ribbonfarm.com/2019/12/17/mediating-consent/.
13. Virginia Eubanks, *Automating Inequality: How High-Tech Tools Profile, Police, and Punish the Poor* (New York: St. Martin's Press, 2017).
14. Brianna Posadas, "How Strategic Is Chicago's 'Strategic Subjects List'? Upturn Investigates," Medium, June 26, 2017, https://medium.com/equal -future/how-strategic-is-chicagos-strategic-subjects-list-upturn-investigates -9e5b4b235a7c.
15. See Martha Poon, "Corporate Capitalism and the Growing Power of Big Data: Review Essay," *Science, Technology, & Human Values* 41, no. 6 (2016): 1088–1108.
16. Whittaker, "The Steep Cost of Capture"; Rodrigo Ochigame, "The Invention of 'Ethical AI': How Big Tech Manipulates Academia to Avoid Regulation," *The Intercept* (blog), December 20, 2019, https://theintercept .com/2019/12/20/mit-ethical-ai-artificial-intelligence/; Thao Phan et al., "Economies of Virtue: The Circulation of 'Ethics' in Big Tech," *Science as Culture*, November 4, 2021, 1–15, https://doi.org/10.1080/09505431.2021 .1990875; Matthew Le Bui and Safiya Umoja Noble, "We're Missing a Moral Framework of Justice in Artificial Intelligence," *The Oxford Handbook of Ethics of AI*, Markus Dirk Dubber, Frank Pasquale, Sunit Das, eds. (Oxford: Oxford University Press, 2020), https://doi.org/10.1093/oxfordhb /9780190067397.013.9.
17. For private ordering, see Jennifer S Fan, "Employees as Regulators: The New Private Ordering in High Technology Companies," *Utah Law Review*, no. 5 (2019): 55.
18. "Principles for Accountable Algorithms and a Social Impact Statement for Algorithms :: FAT ML," accessed October 1, 2018, http://www.fatml .org/resources/principles-for-accountable-algorithms.
19. Important critical studies incorporating historical approaches include Wendy Hui Kyong Chun and Alex Barnett, *Discriminating Data: Correlation, Neighborhoods, and the New Politics of Recognition* (Cambridge, MA: MIT Press, 2021); Justin Joque, *Revolutionary Mathematics: Artificial Intelligence, Statistics and the Logic of Capitalism* (New York: Verso, 2022); Kate Crawford, *Atlas of AI: Power, Politics, and the Planetary Costs of Artificial Intelligence* (New Haven, CT: Yale University Press, 2021); Meredith Broussard, *Artificial Unintelligence: How Computers Misunderstand the World*

(Cambridge, MA: MIT Press, 2018). For "big data," see the pioneering Rob Kitchin, *The Data Revolution: Big Data, Open Data, Data Infrastructures & Their Consequences* (Los Angeles: SAGE Publications, 2014).

20. Melvin Kranzberg, "Technology and History: 'Kranzberg's Laws,' " *Technology and Culture* 27, no. 3 (1986): 547–48.

21. Enrico Coiera, "The Fate of Medicine in the Time of AI," *The Lancet* 392, no. 10162 (December 1, 2018): 2331, https://doi.org/10.1016/S0140-6736(18)31925-1.

22. For history as a powerful tool in teaching ethics, see R. R. Kline, "Using History and Sociology to Teach Engineering Ethics," *IEEE Technology and Society Magazine* 20, no. 4 (2001): 13–20, https://doi.org/10.1109/44.974503.

23. For an excellent survey of different moments of data accumulation and analysis in the US, see Dan Bouk, "The History and Political Economy of Personal Data over the Last Two Centuries in Three Acts," *Osiris* 32 (2017): 85–106; Martha Hodes, "Fractions and Fictions in the United States Census of 1890," in *Haunted by Empire: Geographies of Intimacy in North American History*, ed. Ann Laura Stoler (Durham, NC: Duke University Press, 2006), 240–70; Simone Browne, *Dark Matters: On the Surveillance of Blackness* (Durham, NC: Duke University Press, 2015); Khalil Gibran Muhammad, *The Condemnation of Blackness: Race, Crime, and the Making of Modern Urban America* (Cambridge, MA: Harvard University Press, 2010).

24. Sarah E. Igo, *The Averaged American: Surveys, Citizens, and the Making of a Mass Public* (Cambridge, MA: Harvard University Press, 2007); Emmanuel Didier, *America by the Numbers: Quantification, Democracy, and the Birth of National Statistics* (Cambridge, MA: MIT Press, 2020); Daniel B. Bouk, *How Our Days Became Numbered: Risk and the Rise of the Statistical Individual* (London: University of Chicago Press, 2015); Emily Klancher Merchant, *Building the Population Bomb* (New York: Oxford University Press, 2021). More generally, see the classic studies Geoffrey C. Bowker and Susan Leigh Star, *Sorting Things Out* (Cambridge, MA: MIT Press, 1999) and Wendy Nelson Espeland and Michael Sauder, "Rankings and Reactivity: How Public Measures Recreate Social Worlds," *American Journal of Sociology* 113, no. 1 (July 1, 2007): 1–40, https://doi.org/10.1086/517897.

25. Caitlin Rosenthal, *Accounting for Slavery: Masters and Management* (Cambridge, MA: Harvard University Press, 2018).

26. Theodore M. Porter, *Trust in Numbers: The Pursuit of Objectivity in Science and Public Life* (Princeton, NJ: Princeton University Press, 1995).

27. See the critique in Frank Pasquale, *The Black Box Society: The Secret Algorithms That Control Money and Information* (Cambridge, MA: Harvard University Press, 2015).

28. Martha Poon stresses that the apparent inevitability of these practices is what must be explained, not taken for given: "the details of how scoring systems are made, how they connect, co-ordinate, and interact, and most of all, how they evolve, should matter in how they have reformatted and reassembled the consumer credit industry through risk calculation." Martha Poon, "Scorecards as Devices for Consumer Credit: The Case of Fair, Isaac & Company Incorporated," *The Sociological Review* 55, no. 2_suppl (October 2007): 288, https://doi.org/10.1111/j.1467-954X.2007.00740.x.

CHAPTER 2: SOCIAL PHYSICS AND *L'HOMME MOYEN*

1. Quoted in Karl Pearson, *The Life, Letters and Labours of Francis Galton* (Cambridge, UK: University Press, 1914), vol. 2, 418, http://archive.org/details/b29000695_0002.
2. Nightingale to William Farr, 23.2.1874, in *Florence Nightingale on Society and Politics, Philosophy, Science, Education and Literature: Collected Works of Florence Nightingale*, Volume 5, ed. Lynn McDonald (Waterloo, ON: Wilfrid Laurier University Press, 2003), 39.
3. Ian Hacking, *The Taming of Chance* (Cambridge, UK: Cambridge University Press, 1990), 106.
4. Joseph Lottin, *Quetelet, Statisticien et Sociologue* (Louvain: Institut supérieur de philosophie, 1912), 52; Theodore M. Porter, *The Rise of Statistical Thinking, 1820–1900* (Princeton, NJ: Princeton University Press, 1986), 47.
5. David Aubin, "Principles of Mechanics That Are Susceptible of Application to Society: An Unpublished Notebook of Adolphe Quetelet at the Root of His Social Physics," *Historia Mathematica* 41, no. 2 (May 1, 2014): 209, 216, https://doi.org/10.1016/j.hm.2014.01.001.
6. Kevin Donnelly, *Adolphe Quetelet, Social Physics and the Average Men of Science, 1796–1874* (Routledge, 2015), 73, https://doi.org/10.4324/9781315653662.
7. Translated in Paul F. Lazarsfeld, "Notes on the History of Quantification in Sociology—Trends, Sources and Problems," *Isis* 52, no. 2 (1961): 293.
8. See Morgane Labbé, "L'arithmétique politique en Allemagne au début du 19e siècle: réceptions et polémiques," *Journal Electronique d'Histoire des Probabilités et de la Statistique* 4, no. 1 (2008): 7.
9. Peggy Noonan, "They've Lost That Lovin' Feeling," *Wall Street Journal*, accessed November 20, 2012, http://online.wsj.com/article/SB10001424053111904800304576474620336602248.html.
10. For powerful examinations of developments in the seventeenth century forward, see, among an extensive literature, Jacqueline Wernimont, *Numbered Lives: Life and Death in Quantum Media* (Cambridge, MA: MIT Press, 2018), esp. ch. 2; Andrea Rusnock, "Quantification, Precision, and Accuracy: Determinations of Population in the Ancien Régime," in *Values of Precision*, ed. M. Norton Wise (Princeton, NJ: Princeton University Press, 1995), 17–38; William Deringer, *Calculated Values: Finance, Politics, and the Quantitative Age* (Cambridge, MA: Harvard University Press, 2018), who stresses the nongovernmental origins of much earlier numerical thinking.
11. Lisa Gitelman, ed., *Raw Data Is an Oxymoron.* (Cambridge, MA: MIT Press, 2013). Several major collections of essays develop these ideas across time and the globe: Elena Aronova, Christina von Oertzen, and David Sepkoski, eds., *Data Histories*, Osiris 32 (Chicago: University of Chicago Press, 2017); Soraya de Chadarevian and Theodore M. Porter, *Histories of Data and the Database*, vol. 48, no. 5, *Historical Studies in the Natural Sciences*, 2018; see also Amelia Acker, "Toward a Hermeneutics of Data," *Annals of the History of Computing, IEEE* 37, no. 3 (2015): 70–75.
12. Hacking, *The Taming of Chance*, 2.

13. For a nuanced account of China, see Tong Lam, *A Passion for Facts: Social Surveys and the Construction of the Chinese Nation State, 1900–1949.* (Berkeley: University of California Press, 2011), ch. 1.

14. Jacqueline Wernimont, *Numbered Lives,* 28; for the labor of collecting this data, see Deborah E. Harkness, "A View from the Streets: Women and Medical Work in Elizabethan London," *Bulletin of the History of Medicine* 82, no. 1 (2008): 52–85.

15. Victor L. Hilts, "Aliis Exterendum, or, the Origins of the Statistical Society of London," *Isis* 69, no. 1 (March 1978): 21–43, https://doi.org/10.1086/351931.

16. The focus on argument is central in William Deringer, *Calculated Values.*

17. See David Sepkoski and Marco Tamborini, " 'An Image of Science': Cameralism, Statistics, and the Visual Language of Natural History in the Nineteenth Century," *Historical Studies in the Natural Sciences* 48, no. 1 (February 1, 2018): 56–109, https://doi.org/10.1525/hsns.2018.48.1.56; Deringer, *Calculated Values.*

18. See Jean-Guy Prévost and Jean-Pierre Beaud, *Statistics, Public Debate and the State, 1800–1945: A Social, Political and Intellectual History of Numbers* (Routledge, 2016), 3.

19. Quetelet's international work is a major theme of Donnelly, *Adolphe Quetelet, Social Physics and the Average Men of Science, 1796–1874.*

20. Adolphe Quetelet, *A Treatise on Man and the Development of His Faculties* (Edinburgh: W. and R. Chambers, 1842), 6, http://archive.org/details/treatise onmandev00quet.

21. Quetelet, 6.

22. David Aubin, "On the Epistemic and Social Foundations of Mathematics as Tool and Instrument in Observatories, 1793–1846," in *Mathematics as a Tool,* ed. Johannes Lenhard and Martin Carrier, vol. 327 (Cham, Switzerland: Springer International Publishing, 2017), 290–91, https://doi.org/10 .1007/978-3-319-54469-4_10; Porter, *The Rise of Statistical Thinking, 1820-1900,* 42; Donnelly, *Adolphe Quetelet, Social Physics and the Average Men of Science, 1796–1874,* 111–12. Aubin stresses the techniques of observing data; Donnelly emphasizes Quetelet's increasing ability to get data from sources across Europe.

23. Hacking, *The Taming of Chance,* 109.

24. Adolphe Quetelet, *Recherches Statistiques* (Brussels: M. Hayez, 1844), 54.

25. Hacking, *The Taming of Chance,* 107.

26. Quetelet, *A Treatise on Man,* 5.

27. Quetelet, 6.

28. Quetelet, 6.

29. Quetelet, 6.

30. Quetelet, 6.

31. Hacking, *The Taming of Chance,* 108.

32. Margaret Thatcher, Interview for Woman's Own, 23.9.1987, https://www .margaretthatcher.org/document/106689.

33. Porter, *The Rise of Statistical Thinking, 1820–1900,* 55.

34. Quetelet, *A Treatise on Man,* 7.

35. Porter, *The Rise of Statistical Thinking, 1820–1900,* 46.

36. Porter, 104.

37. Hacking, *Taming of Chance,* 108.

38. Adrian Wooldridge, *Measuring the Mind: Education and Psychology in England, c. 1860–c. 1990* (New York: Cambridge University Press, 1994), 74.
39. Quoted in Pearson, *The Life, Letters and Labours of Francis Galton*, v. 2, 419.
40. Quoted in Pearson, v. 2, 419.

CHAPTER 3: THE STATISTICS OF THE DEVIANT

1. Florence Nightingale, *Notes on Matters Affecting the Health, Efficiency and Hospital Administration of the British Army* (London: Harrison and Sons, 1858), 518.
2. Francis Galton, "Heredity Talent And Character," *Macmillan's Magazine* 12 (1865): 166.
3. Galton explained that Darwin "made a marked epoch in my own mental development, as it did in that of human thought generally." Sir Francis Galton, *Memories of My Life* (New York: Dutton, 1909), 287.
4. Galton, "Heredity Talent And Character," 157.
5. Galton, 165.
6. Chris Renwick, "From Political Economy to Sociology: Francis Galton and the Social-Scientific Origins of Eugenics," *The British Journal for the History of Science* 44, no. 3 (September 2011): 352, https://doi.org/10.1017/S000 7087410001524.
7. Francis Galton, *Hereditary Genius: An Inquiry into Its Laws and Consequences* (London: Macmillan, 1869), 14, http://archive.org/details/hereditary genius1869galt.
8. Ross, quoted in Thomas C. Leonard, *Illiberal Reformers: Race, Eugenics, and American Economics in the Progressive Era* (Princeton, NJ: Princeton University Press, 2016), 110.
9. For connections between this eugenics and contemporary data practices, see Chun and Barnett, *Discriminating Data*, ch. 1.
10. Alain Desrosières, *The Politics of Large Numbers: A History of Statistical Reasoning* (Cambridge, MA: Harvard University Press, 1998), 113; Stephen M. Stigler, *The History of Statistics: The Measurement of Uncertainty Before 1900* (Cambridge, MA: The Belknap Press of Harvard University Press, 1986), 271.
11. Francis Galton, "Typical Laws of Heredity," *Royal Institution of Great Britain. Notices of the Proceedings at the Meetings of the Members* 8 (February 16, 1877): 291.
12. Francis Galton, *Anthropometric Laboratory; Arranged by Francis Galton, FRS, for the Determination of Height, Weight, Span, Breathing Power, Strength of Pull and Squeeze, Quickness of Blow, Hearing, Seeing, Colour-Sense, and Other Personal Data* (London: William Clowes, 1884), 3, http://archive.org/details/b30579132.
13. Galton, 4.
14. Kurt Danziger, *Constructing the Subject: Historical Origins of Psychological Research* (Cambridge, UK: Cambridge University Press, 1990), 57.
15. Danziger, 77.
16. Danziger, 110.
17. Porter, *The Rise of Statistical Thinking, 1820–1900*, 311.
18. Porter, 304–5. "In another sense, Pearson was a true follower of Quetelet. Both agreed on the universality of numbers and on the absence of discontinuity. Both maintained that the task of science was not to chart a bold new

course, but to study the laws of social development so that scientific policy might affirm them and remove all obstacles to their attainment."

19. Theodore M. Porter, *Karl Pearson: The Scientific Life in a Statistical Age* (Princeton, NJ: Princeton University Press, 2004), 261.
20. Karl Pearson, "On the Laws of Inheritance in Man: II. On the Inheritance of the Mental and Moral Characters in Man, and Its Comparison with the Inheritance of the Physical Characters," *Biometrika* 3, no. 2/3 (1904): 136, https://doi.org/10.2307/2331479.
21. M. Eileen Magnello, "The Non-Correlation of Biometrics and Eugenics: Rival Forms of Laboratory Work in Karl Pearson's Career at University College London, Part 1," *History of Science* 37, no. 1 (March 1, 1999): 79–106, https://doi.org/10.1177/007327539903700103; M. Eileen Magnello, "The Non-Correlation of Biometrics and Eugenics: Rival Forms of Laboratory Work in Karl Pearson's Career at University College London, Part 2," *History of Science* 37, no. 2 (June 1, 1999): 123–50.
22. Porter, *Karl Pearson: The Scientific Life in a Statistical Age*, 263.
23. Pearson, "On the Laws of Inheritance in Man," 136. On women as computers, see Jennifer S. Light, "When Computers Were Women," *Technology and Culture* 40, no. 3 (1999): 455–83.
24. David Alan Grier, *When Computers Were Human* (Princeton, NJ: Princeton University Press, 2005), 111.
25. Quoted in Grier, 117.
26. Pearson, *The Life, Letters and Labours of Francis Galton*, IIIA: 305.
27. Pearson, "On the Laws of Inheritance in Man," 159.
28. Alice Lee and Karl Pearson, "Data for the Problem of Evolution in Man. VI. A First Study of the Correlation of the Human Skull," *Philosophical Transactions of the Royal Society of London. Series A, Containing Papers of a Mathematical or Physical Character* 196 (1901): 259.
29. Karl Pearson, "On the Inheritance of the Mental and Moral Characters in Man, and Its Comparison with the Inheritance of the Physical Characters," *The Journal of the Anthropological Institute of Great Britain and Ireland* 33 (1903): 207, https://doi.org/10.2307/2842809.
30. Karl Pearson and Margaret Moul, "The Problem of Alien Immigration into Great Britain, Illustrated by an Examination of Russian and Polish Jewish Children," *Annals of Eugenics* 1, no. 1 (1925): 7, https://doi.org/10.1111/j.1469-1809.1925.tb02037.x.
31. Karl Pearson, *The Chances of Death, and Other Studies in Evolution* (London, New York: E. Arnold, 1897), 104, http://archive.org/details/cu31924097311579. Quoted in Porter, *Karl Pearson: The Scientific Life in a Statistical Age*, 267.
32. Karl Pearson and Ethel M. Elderton, *A Second Study of the Influence of Parental Alcoholism on the Physique and Ability of the Offspring: Being a Reply to Certain Medical Critics of the First Memoir and an Examination of the Rebutting Evidence Cited by Them* (London, Dulau and Co., 1910), 34, http://archive.org/details/secondstudyofinf00pear; discussed in P. C. Mahalanobis, "Karl Pearson, 1857–1936," *Sankhyā: The Indian Journal of Statistics* 2, no. 4 (1936): 368; see also Donald A. MacKenzie, *Statistics in Britain, 1865–1930: The Social Construction of Scientific Knowledge* (Edinburgh: Edinburgh University Press, 1981), 139.
33. Michel Armatte, "Invention et intervention statistiques. Une conférence

exemplaire de Karl Pearson (1912)," *Politix. Revue des sciences sociales du politique* 7, no. 25 (1994): 30, https://doi.org/10.3406/polix.1994.1823.

34. Porter, *The Rise of Statistical Thinking, 1820–1900*, 298.

35. Pearson, *The Life, Letters and Labours of Francis Galton*, IIIa:57.

36. Robert A. Nye, "The Rise and Fall of the Eugenics Empire: Recent Perspectives on the Impact of Biomedical Thought in Modern Society," *The Historical Journal* 36, no. 3 (September 1993): 695, https://doi.org/10.1017/S00 18246X00014369.

37. Brajendranath Seal, "Meaning of Race, Tribe, Nation," in *Papers on Inter-Racial Problems, Communicated to the First Universal Races Congress, Held at the University of London, July 26–29, 1911*, ed. Gustav Spiller (London: P. S. King & Son; Boston, The World's Peace Foundation, 1911), 1, http://archive.org/details/papersoninterrac00univiala; our remarks much indebted to Projit Bihari Mukharji, "The Bengali Pharaoh: Upper-Caste Aryanism, Pan-Egyptianism, and the Contested History of Biometric Nationalism in Twentieth-Century Bengal," *Comparative Studies in Society and History* 59, no. 2 (April 2017): 450, https://doi.org/10.1017/S001041751700010X. For his broader statistical program, see Theodora Dryer, "Designing Certainty: The Rise of Algorithmic Computing in an Age of Anxiety 1920–1970" (PhD Thesis, UC San Diego, 20194), 157–162; and see the forthcoming work Sananda Sahoo, "Multiple lives of Mahalanobis' biometric data travel as biovalue to India's welfare state" (under review).

38. Seal, "Meaning of Race, Tribe, Nation," 2.

39. Seal, 3.

40. Nikhil Menon, " 'Fancy Calculating Machine': Computers and Planning in Independent India," *Modern Asian Studies* 52, no. 2 (March 2018): 421–57, https://doi.org/10.1017/S0026749X16000135; Sandeep Mertia, "Did Mahalanobis Dream of Androids?," in *Lives of Data: Essays on Computational Cultures from India*, ed. Sandeep Mertia and Ravi Sundaram (Amsterdam: Institute of Network Cultures, 2020), 26–33.

41. For this idea, see Projit Bihari Mukharji, "Profiling the Profiloscope: Facialization of Race Technologies and the Rise of Biometric Nationalism in Inter-War British India," *History and Technology* 31, no. 4 (October 2, 2015): 392, https://doi.org/10.1080/07341512.2015.1127459. "Profiling the Profiloscope allows us then to glimpse an important moment in the development of biometric technologies in South Asia in particular, but also more generally. It shows us that in the interwar period, nationalists rather than the weakened colonial state developed biometric technologies."

42. P. C. Mahalanobis, "Analysis of Race-Mixture in Bengal," *Journal of the Asiatic Society of Bengal* 23 (1927): 323.

43. P. C. Mahalanobis et al., "Anthropometric Survey of the United Provinces, 1941: A Statistical Study," *Sankhyā: The Indian Journal of Statistics* 9, no. 2/3 (1949): 168.

44. Mahalanobis et al., 180.

45. W. E. Burghardt Du Bois, "A Summary of the Main Conclusions of the Papers Presented to the First Universal Races Conference," Series 1, Box 007, Special Collections and University Archives, University of Massachusetts Amherst Libraries, https://www.digitalcommonwealth.org/search/commonwealth-oai:h128q9079.

CHAPTER 4: DATA, INTELLIGENCE, AND POLICY

1. Frederick L. Hoffman, *The Race Traits and Tendencies of the American Negro* (Publications of the American Economic Association, 1896), 2, http://archive.org/details/jstor-2560438. For Hoffman, see Daniel B. Bouk, *How Our Days Became Numbered: Risk and the Rise of the Statistical Individual* (London: University of Chicago Press, 2015), 48–52.

2. Hoffman, *The Race Traits and Tendencies of the American Negro*, 312.

3. See Beatrix Hoffman, "Scientific Racism, Insurance, and Opposition to the Welfare State: Frederick L. Hoffman's Transatlantic Journey," *The Journal of the Gilded Age and Progressive Era* 2, no. 2 (April 2003): 150–90, https://doi.org/10.1017/S1537781400002450.

4. W. E. Burghardt Du Bois, "Review of 'Race Traits and Tendencies of the American Negro,' " *The Annals of the American Academy of Political and Social Science* 9, no. 1 (1897): 129.

5. Ayah Nurddin, "The Black Politics of Eugenics," *Nursing Clio* (blog), June 1, 2017, https://nursingclio.org/2017/06/01/the-black-politics-of-eugenics/.

6. George M. Fredrickson, *The Black Image in the White Mind: The Debate on Afro-American Character and Destiny, 1817–1914* (Scranton, PA: Distributed by Harper & Row, 1987), 249.

7. Khalil Gibran Muhammad, *The Condemnation of Blackness: Race, Crime, and the Making of Modern Urban America* (Cambridge, MA: Harvard University Press, 2010), 5.

8. Desrosières, *The Politics of Large Numbers*, 139.

9. T. S. Simey, *Charles Booth, Social Scientist* (London, 1960), 48, http://hdl .handle.net/2027/uc1.b3620533.

10. Charles Booth, *The Aged Poor in England and Wales* (London: Macmillan and Co., 1894), 423, http://archive.org/details/agedpoorinengla00bootgoog.

11. Compare Stigler, *The History of Statistics*, 354.

12. Karl Pearson, *The Grammar of Science*, 3rd ed. (London: Adam & Charles Black, 1911), 157.

13. G. Udny Yule, "On the Theory of Correlation," *Journal of the Royal Statistical Society* 60, no. 4 (December 1897): 812, https://doi.org/10.2307/2979746.

14. G. Udny Yule, "An Investigation into the Causes of Changes in Pauperism in England, Chiefly During the Last Two Intercensal Decades (Part I.)," *Journal of the Royal Statistical Society* 62, no. 2 (1899): 249, https://doi .org/10.2307/2979889.

15. Yule, 250.

16. G. Udny Yule, "On the Correlation of Total Pauperism with Proportion of Out-Relief," *The Economic Journal* 5, no. 20 (1895): 605, https://doi.org/10 .2307/2956650; discussed in C. Terence Mills, *A Statistical Biography of George Udny Yule: A Loafer of the World* (Newcastle upon Tyne: Cambridge Scholars Publisher, 2017), 43.

17. G. Udny Yule, "On the Correlation of Total Pauperism with Proportion of Out-Relief," *The Economic Journal* 5, no. 20 (1895): 606, https://doi.org/10 .2307/2956650.

18. He went further to note that "detailed knowledge" might give some causal understanding: "Detailed knowledge may occasionally enable one to say 'The pauperism is low here, since the proportion of out-relief is very small,' or perhaps, 'The proportion of out-relief given is large on account of the

high pauperism and other industrial conditions of the union': but such cases will be exceptional and will as a rule only refer to large deviations from the mean." Yule, "On the Correlation of Total Pauperism with Proportion of Out-Relief," 1895, 605n2; discussed in Mills, *Statistical Biography*, 46.

19. Yule, "An Investigation into the Causes of Changes in Pauperism in England, Chiefly During the Last Two Intercensal Decades (Part I.)," 251.

20. "There is still a certain chance of error depending on the number of factors correlated both with pauperism and with proportion of out-relief which have been omitted, but obviously this chance of error will be much smaller than before." Yule, 251.

21. Stigler, *The History of Statistics*, 356.

22. Yule, "An Investigation into the Causes of Changes in Pauperism in England, Chiefly During the Last Two Intercensal Decades (Part I.)," 265.

23. Yule, 257n.16.

24. Yule, 277.

25. Yule, "On the Theory of Correlation," 812.

26. Arthur Cecil Pigou, "Memorandum on Some Economic Aspects and Effects of Poor Law Relief," in *Royal Commission on the Poor Laws and Relief of Distress, Appendix, Vol. 9*, Parliamentary Papers for the Session 15 February 1910–28 November 1910, Vol. 49. (London: His Majesty's Stationery Office, 1910), 984–85.

27. Pigou, 986.

28. Pigou, 986.

29. David A. Freedman, "Statistical Models and Shoe Leather," *Sociological Methodology* 21 (1991): 291, https://doi.org/10.2307/270939.

30. Yule, "An Investigation into the Causes of Changes in Pauperism in England, Chiefly During the Last Two Intercensal Decades (Part I.)," 270.

31. Desrosières, *The Politics of Large Numbers*, 140.

32. Shivrang Setlur, "Searching for South Asian Intelligence: Psychometry in British India, 1919–1940," *Journal of the History of the Behavioral Sciences* 50, no. 4 (2014): 359–75, https://doi.org/10.1002/jhbs.21692.

33. Charles Spearman, " 'General Intelligence,' Objectively Determined and Measured," *The American Journal of Psychology* 15, no. 2 (April 1904): 277, https://doi.org/10.2307/1412107 (our italics).

34. Adrian Wooldridge, *Measuring the Mind: Education and Psychology in England, c. 1860–c. 1990* (New York: Cambridge University Press, 1994), 74.

35. Charles Spearman, *The Nature of "Intelligence" and the Principles of Cognition* (London: Macmillan, 1923), 355; quoted in Stephen Jay Gould, *The Mismeasure of Man* (New York: Norton, 1996), 293.

36. Charles Spearman, *The Abilities of Man: Their Nature and Measurement* (New York: The Macmillan Company, 1927), 379; quoted in Gould, *The Mismeasure of Man*, 301, 302.

37. Spearman, *The Abilities of Man; Their Nature and Measurement*, 380.

38. John Carson, *The Measure of Merit: Talents, Intelligence, and Inequality in the French and American Republics, 1750–1940* (Princeton, NJ: Princeton University Press, 2007), 183–93 provides an excellent survey of debates about measuring intelligence.

39. Karl Pearson and Margaret Moul, "The Mathematics of Intelligence. The Sampling Errors in the Theory of a Generalised Factor," *Biometrika* 19, no.

3/4 (1927): 291, https://doi.org/10.2307/2331962. See Theodore M. Porter, *Karl Pearson: The Scientific Life in a Statistical Age* (Princeton, NJ: Princeton University Press, 2004), 270.

40. Carson, *The Measure of Merit*, 159.

41. Richard J. Herrnstein and Charles A. Murray, *The Bell Curve: Intelligence and Class Structure in American Life* (New York: Simon & Schuster, 1996).

42. Colin Koopman, *How We Became Our Data: A Genealogy of the Informational Person* (Chicago: The University of Chicago Press, 2019). Enslaved people had been subject to careful accounting in the nineteenth century, see Caitlin Rosenthal, *Accounting for Slavery: Masters and Management* (Cambridge, MA: Harvard University Press, 2018), esp. ch. 2; Simone Browne, *Dark Matters: On the Surveillance of Blackness* (Durham, NC: Duke University Press, 2015).

43. Wangui Muigai, in Projit Bihari Mukharji et al., "A Roundtable Discussion on Collecting Demographics Data," *Isis* 111, no. 2 (June 2020): 320, https://doi.org/10.1086/709484.

44. Wangui Muigai, in Mukharji et al., 320. For the continuing salience of assumptions of gender categories, see Mar Hicks, "Hacking the Cis-Tem," *IEEE Annals of the History of Computing* 41, no. 1 (January 2019): 20–33, https://doi.org/10.1109/MAHC.2019.2897667.

45. Quoted in Sandeep Mertia, "Did Mahalanobis Dream of Androids?," in *Lives of Data: Essays on Computational Cultures from India*, ed. Sandeep Mertia and Ravi Sundaram (Amsterdam: Institute of Network Cultures, 2020), 31.

46. Mertia, 31.

47. Emmanuel Didier, *America by the Numbers: Quantification, Democracy, and the Birth of National Statistics* (Cambridge, MA: MIT Press, 2020), 11.

48. J. Adam Tooze, *Statistics and the German State, 1900-1945: The Making of Modern Economic Knowledge* (New York: Cambridge University Press, 2001), 24

49. Arunabh Ghosh, *Making It Count: Statistics and Statecraft in the Early People's Republic of China* (Princeton: Princeton University Press, 2020), 283.

50. Tooze, *Statistics and the German State*, 28.

51. John Koren and Edmund Ezra Day, *The History of Statistics, Their Development and Progress in Many Countries; in Memoirs to Commemorate the Seventy Fifth Anniversary of the American Statistical Association* (New York: Pub. for the American Statistical Association by the Macmillan Company of New York, 1918), 25–26, http://archive.org/details/cu31924013894997.

52. John Stuart Mill, *Principles of Political Economy: With Some of Their Applications to Social Philosophy* (London: J. W. Parker, 1848), 375.

53. Kevin Bird, "Still Not in Our Genes: Resisting the Narrative Around GWAS," *Science for the People Magazine* 23, no. 3 (February 5, 2021), https://magazine.scienceforthepeople.org/vol23-3-bio-politics/genetic-basis-genome-wide-association-studies-risk/.

54. Kelly Miller, *A Review of Hoffman's* Race Traits and Tendencies of the American Negro, American Negro Academy. Occasional Papers, no. 1 (Washington, DC: The Academy, 1897), 35, https://catalog.hathitrust.org/Record/100788175.

CHAPTER 5: DATA'S MATHEMATICAL BAPTISM

1. Joan Fisher Box, "Guinness, Gosset, Fisher, and Small Samples," *Statistical Science* 2, no. 1 (1987): 48.
2. Student, "On Testing Varieties of Cereals," *Biometrika* 15, no. 3/4 (1923): 271, https://doi.org/10.2307/2331868.
3. Student, "The Probable Error of a Mean," *Biometrika* 6, no. 1 (1908): 2, https://doi.org/10.2307/2331554.
4. Box, "Guinness, Gosset, Fisher, and Small Samples." Box wonderfully evokes Gosset's world and work.
5. E. S. Pearson, " 'Student' as Statistician," *Biometrika* 30, no. 3/4 (January 1939): 215–16, https://doi.org/10.2307/2332648.
6. Box, "Guinness, Gosset, Fisher, and Small Samples," 49.
7. Donald A. MacKenzie, *Statistics in Britain, 1865–1930: The Social Construction of Scientific Knowledge* (Edinburgh: Edinburgh University Press, 1981), 111f; for a readable appreciation of Gosset, see especially Stephen Thomas Ziliak and Deirdre N. McCloskey, *The Cult of Statistical Significance: How the Standard Error Costs Us Jobs, Justice, and Lives* (Ann Arbor: University of Michigan Press, 2008).
8. Nan M. Laird, "A Conversation with F. N. David," *Statistical Science* 4, no. 3 (August 1989): 238, https://doi.org/10.1214/ss/1177012487.
9. Ronald Aylmer Fisher, *Statistical Methods for Research Workers* (London: Oliver and Boyd, 1925), vii.
10. See especially Giuditta Parolini, "The Emergence of Modern Statistics in Agricultural Science: Analysis of Variance, Experimental Design and the Reshaping of Research at Rothamsted Experimental Station, 1919–1933," *Journal of the History of Biology* 48, no. 2 (May 2015): 301–35, https://doi.org/10.1007/s10739-014-9394-z.
11. Box, "Guinness, Gosset, Fisher, and Small Samples," 51.
12. E. L. Lehmann, *Fisher, Neyman, and the Creation of Classical Statistics* (New York: Springer, 2011), 12.
13. R. A. Fisher and W. A. Mackenzie, "Studies in Crop Variation. II. The Manurial Response of Different Potato Varieties," *The Journal of Agricultural Science* 13 (1923): 469.
14. Fisher, *Statistical Methods for Research Workers*, vii.
15. Fisher, 4.
16. Ronald Aylmer Fisher, *The Design of Experiments* (London: Oliver and Boyd, 1935), 15–16.
17. Fisher, 49. For the development of randomization, see Nancy S. Hall, "R. A. Fisher and His Advocacy of Randomization," *Journal of the History of Biology* 40, no. 2 (June 1, 2007): 295–325, https://doi.org/10.1007/s10739-006-9119-z.
18. Epstein, *Impure Science: AIDS, Activism, and the Politics of Knowledge* (Berkeley: University of California Press, 1996).
19. Stephen T. Ziliak, "W.S. Gosset and Some Neglected Concepts in Experimental Statistics: Guinnessometrics II*," *Journal of Wine Economics* 6, no. 2 (ed 2011): 252–77, https://doi.org/10.1017/S1931436100001632.
20. Fisher, *The Design of Experiments*, 10.
21. Fisher, 10.
22. Ronald Aylmer Fisher, "Some Hopes of a Eugenicist," in *Collected Papers of R. A. Fisher*, vol. 1 (Adelaide: University of Adelaide, 1971), 78. For Fisher

as a eugenicist, see Alex Aylward, "R.A. Fisher, Eugenics, and the Campaign for Family Allowances in Interwar Britain," *The British Journal for the History of Science* 54, no. 4 (December 2021): 485–505, https://doi.org/10.1017/S0007087421000674.

23. Fisher, "Some Hopes of a Eugenicist," 79.

24. Ronald Fisher, "Statistical Methods and Scientific Induction," *Journal of the Royal Statistical Society: Series B (Methodological)* 17, no. 1 (January 1, 1955): 75, https://doi.org/10.1111/j.2517-6161.1955.tb00180.x.

25. J. Neyman and E. S. Pearson, "On the Problem of the Most Efficient Tests of Statistical Hypotheses," *Philosophical Transactions of the Royal Society of London. Series A, Containing Papers of a Mathematical or Physical Character* 231 (1933): 291.

26. Theodora Dryer, "Designing Certainty: The Rise of Algorithmic Computing in an Age of Anxiety 1920–1970" (PhD Thesis, UC San Diego, 2019), 81.

27. Constance Reid, *Neyman* (New York: Springer, 1998), 24–25, 48. She's describing a manuscript whose first part "is concerned with principles which justify the use of abstract mathematical theory in studies of natural phenomena, especially in the domain of agricultural experimentation."

28. J. Neyman, " 'Inductive Behavior' as a Basic Concept of Philosophy of Science," *Revue de l'Institut International de Statistique / Review of the International Statistical Institute* 25, no. 1/3 (1957): 8, https://doi.org/10.2307/1401671.

29. Karl Pearson, *The Grammar of Science* (London: Walter Scott ; New York: Charles Scribner's Sons, 1892), 72, http://archive.org/details/grammarofscience00pearrich.

30. Gosset to Egon Pearson, 11.5.1926, in Pearson, " 'Student' as Statistician," 243. See Lehmann, *Fisher, Neyman, and the Creation of Classical Statistics*, 7.

31. Gosset to Egon Pearson, 11.5.1926, in Pearson, " 'Student' as Statistician," 242.

32. Jerzy Neyman, *A Selection of Early Statistical Papers of J. Neyman.* (Berkeley: University of California Press, 1967), 352.

33. Lehmann, *Fisher, Neyman, and the Creation of Classical Statistics*, 37 (our italics).

34. Lehmann, *Fisher, Neyman, and the Creation of Classical Statistics*; Gerd Gigerenzer, ed., *The Empire of Chance: How Probability Changed Science and Everyday Life* (Cambridge, UK: Cambridge University Press, 1989). For the subsequent big picture in the social sciences, see Hunter Heyck, *Age of System* (Baltimore: Johns Hopkins University Press, 2015), ch. 4.

35. Neyman, *A Selection of Early Statistical Papers of J. Neyman*, 352.

36. Ronald Aylmer Fisher, "Scientific Thought and the Refinement of Human Reasoning," *Journal of the Operations Research Society of Japan* 3 (1960): 3. Justin Joque, *Revolutionary Mathematics: Artificial Intelligence, Statistics and the Logic of Capitalism* (New York: Verso, 2022), ch. 5 offers a rich philosophical reading of this contest.

37. Ronald Aylmer Fisher, *Statistical Methods and Scientific Inference* (Edinburgh: Oliver and Boyd, 1956), 7, http://archive.org/details/statisticalmetho0000fish.

38. Gerd Gigerenzer and Julian N. Marewski, "Surrogate Science: The Idol of a Universal Method for Scientific Inference," *Journal of Management* 41, no. 2 (February 1, 2015): 421–40, https://doi.org/10.1177/0149206314547522.

39. Snedecor's *Statistical Methods* was a key vector.
40. Christopher Phillips, "Inference Rituals: Algorithms and the History of Statistics," in *Algorithmic Modernity: Mechanizing Thought and Action, 1500–2000*, ed. Massimo Mazzotti and Morgan Ames (Oxford, UK: Oxford University Press, forthcoming).
41. Theodore M. Porter, *Trust in Numbers: The Pursuit of Objectivity in Science and Public Life* (Princeton, NJ: Princeton University Press, 1995), 206.
42. Porter, 206.
43. Gigerenzer, *The Empire of Chance*, 106.
44. W. Allen Wallis, "The Statistical Research Group, 1942–1945," *Journal of the American Statistical Association* 75, no. 370 (June 1, 1980): 321, https://doi.org/10.1080/01621459.1980.10477469.
45. Judy L. Klein, "Economics for a Client: The Case of Statistical Quality Control and Sequential Analysis," *History of Political Economy* 32, no. Suppl. 1 (2000): 25–70; Nicola Giocoli, "From Wald to Savage: Homo Economicus Becomes a Bayesian Statistician," *Journal of the History of the Behavioral Sciences* 49, no. 1 (2013): 63–95, https://doi.org/10.1002/jhbs.21579.
46. [Mina Rees], description of mathematics program of ONR, 9/27/1946, Hotelling Papers, Box 18, ONR Contract and Renewals, Columbia University Special Collections.
47. Harold Hotelling, "The Place of Statistics in the University (with Discussion)," in *Proceedings of the [First] Berkeley Symposium on Mathematical Statistics and Probability* (Berkeley, CA, The Regents of the University of California, 1949), 23, https://projecteuclid.org/euclid.bsmsp/1166219196.
48. Jerzy Neyman, ed., *Proceedings of the [First] Berkeley Symposium on Mathematical Statistics and Probability* (Berkeley, CA: The Regents of the University of California, 1949), https://projecteuclid.org/euclid.bsmsp/1166219194.
49. Hotelling, "The Place of Statistics in the University (with Discussion)," 23.
50. John W. Tukey, "The Future of Data Analysis," *The Annals of Mathematical Statistics* 33, no. 1 (1962): 6.

CHAPTER 6: DATA AT WAR

1. Juanita Moody, Oral History, interview by Jean Lichty et al., June 16, 1994, 26, https://media.defense.gov/2021/Jul/15/2002763502/-1/-1/0/NSA-OH-1994-32-MOODY.PDF.
2. Mark Brown, "Bletchley Discloses Real Intention of 1938 'Shooting Party,' " *The Guardian*, September 18, 2018, sec. World news, https://www.theguardian.com/world/2018/sep/18/bletchley-discloses-real-intention-1938-shooting-party-wapark-r.
3. Howard Campaigne, Oral History, interview by Robert D Farley, June 29, 1983, 15–16, https://www.nsa.gov/portals/75/documents/news-features/declassified-documents/oral-history-interviews/nsa-oh-14-83-campaigne.pdf.
4. David Kenyon, *Bletchley Park and D-Day* (New Haven, CT: Yale University Press, 2019), 236.
5. Eleanor Ireland, Oral History, interview by Janet Abbate, April 23, 2001, https://ethw.org/Oral-History:Eleanor_Ireland.
6. J. Abbate, *Recoding Gender: Women's Changing Participation in Computing* (Cambridge, MA: MIT Press, 2012), 20. See also Mar Hicks, *Programmed*

Inequality: How Britain Discarded Women Technologists and Lost Its Edge in Computing (Cambridge, MA: MIT Press, 2017), chap. 1.

7. Abbate, *Recoding Gender: Women's Changing Participation in Computing*, 22.
8. Abbate, 27. Drawing upon her interview with Ireland, Oral History.
9. Quoted in B. Jack Copeland, ed., *Colossus: The Secrets of Bletchley Park's Codebreaking Computers* (Oxford; New York: Oxford University Press, 2006), 171.
10. Hicks, *Programmed Inequality*, 40–41.
11. Abraham Sinkov, Oral History, interview by Arthur J Zoebelein et al., May 1979, 3–4.
12. Solomon Kullback, Oral History, interview by R. D. Farley and H. F. Schorreck, August 26, 1982, 48.
13. Phillip Rogaway, "The Moral Character of Cryptographic Work" (2015), 1, https://web.cs.ucdavis.edu/~rogaway/papers/moral-fn.pdf.
14. W. J. Holmes, *Double Edged Secrets: U.S. Naval Intelligence Operations in the Pacific During World War II.* (Annapolis, MD: Naval Institute Press, 2012), p. 142.
15. Kenyon, *Bletchley Park and D-Day*, 242–43.
16. The first is equivalent to the null hypothesis for Fisher, or to both the null hypothesis and the competing hypothesis for Neyman and his school.
17. Stephen M. Stigler, "The True Title of Bayes's Essay," *Statistical Science* 28, no. 3 (August 2013): 283–88, https://doi.org/10.1214/13-STS438; Richard Swinburne, "Bayes, God, and the Multiverse," in *Probability in the Philosophy of Religion*, ed. Jake Chandler and Victoria S. Harrison (Oxford, UK: Oxford University Press, 2012), 103–26.
18. Moreover, we may disagree on how many hypotheses should be considered.
19. Ian Taylor, "Alan M. Turing: The Applications of Probability to Cryptography," *ArXiv:1505.04714 [Math]*, May 26, 2015, 3, http://arxiv.org/abs/1505 .04714.
20. For Turing's approach, see Sandy Zabell, "Commentary on Alan M. Turing: The Applications of Probability to Cryptography," *Cryptologia* 36, no. 3 (July 2012): 191–214, https://doi.org/10.1080/01611194.2012.697811.
21. F. T. Leahy, "The Apparent Paradox of Bayes Factors (U)," *NSA Technical Journal* 27, no. 3 (n.d.): 8, 9. Compare Turing's own remarks, Taylor, "Alan M. Turing," 2–3.
22. For a popular history of Bayes and its successes, see S. B. McGrayne, *The Theory That Would Not Die: How Bayes' Rule Cracked the Enigma Code, Hunted Down Russian Submarines, & Emerged Triumphant from Two Centuries of Controversy* (New Haven, CT: Yale University Press, 2011). And for a more philosophical reading, Joque, *Revolutionary Mathematics*, chs. 6–7.
23. As documents were declassified, Good wrote explicitly about the roots of his work during the war. Irving J. Good, "Turing's Anticipation of Empirical Bayes in Connection with the Cryptanalysis of the Naval Enigma," *Journal of Statistical Computation and Simulation* 66, no. 2 (2000): 101–11. His writings appear also in classified NSA journals: Irving J. Good, "A List of Properties of Bayes-Turing Factors," *NSA Technical Journal* 10, no. 2 (1965), https://www.nsa.gov/Portals/70/ documents/news-features/declassified-documents/tech-journals/list-of -properties.pdf.

24. Colin B. Burke, *It Wasn't All Magic: The Early Struggle to Automate Crypt-analysis, 1930s–1960s* (Fort Meade, MD: Center for Cryptological History, NSA, 2002), 277, http://archive.org/details/NSA-WasntAllMagic_2002. For reflections on the need to contend with increasing volumes from the 1960s onward, see Willis Ware, "Report of the Second Computer Study Group, Submitted May 1972," *NSA Technical Journal* 19, no. 1 (1974): 21–63; Joseph Eachus et al., "Growing Up with Computers at NSA (Top Secret Umbra)," *NSA Technical Journal* Special issue (1972): 3–14.

25. Burke, *It Wasn't All Magic*, 265.

26. Frances Allen, interview by Paul Lasewicz, April 16, 2003, 4, https:// amturing.acm.org/allen_history.pdf.

27. Allen, 4–5.

28. Frances Allen, Oral History, interview by Al Kossow, September 11, 2008, 5. Computer History Museum X5006.2009.

29. Burke, *It Wasn't All Magic*, 264.

30. Samuel S. Snyder, "Computer Advances Pioneered by Cryptologic Organizations," *Annals of the History of Computing* 2, no. 1 (1980): 66.

31. Samuel S. Snyder, "ABNER: The ASA Computer, Part II: Fabrication, Operation, and Impact," *NSA Technical Journal*, n.d., 83.

32. For developments in the 1970s and 1980s, still mostly classified, see Thomas R. Johnson, *American Cryptology During the Cold War, 1945–1989, Book IV: Cryptologic Rebirth 1981–1989* (NSA Center for Cryptologic History, 1999), 291–292.

33. redacted, "Multiple Hypothesis Testing and the Bayes Factor (Secret)," *NSA Technical Journal* 16, no. 3 (1971): 63–80, p. 71.

34. F. T. Leahy, "The Apparent Paradox of Bayes Factors (U)," *NSA Technical Journal* 27, no. 3 (n.d.): 7–10, pp. 8, 9. "For there can exist for the cryptographer no assignment of a priori odds (whether ingenious or otherwise) that can adversely affect the usefulness of our computer program."

35. Compare several papers on cluster analysis in NSATJ; work of R51.

36. Mina Rees, "The Federal Computing Machine Program," *Science* 112, no. 2921 (December 22, 1950): 735; for a powerful account of Rees's role in the shaping of support for mathematics, see Alma Steingart, *Axiomatics: Mathematical Thought and High Modernism* (Chicago: University of Chicago Press, forthcoming).

37. Robert W. Seidel, " 'Crunching Numbers': Computers and Physical Research in the AEC Laboratories," *History and Technology* 15, no. 1–2 (September 1, 1998): 54, https://doi.org/10.1080/07341519808581940.

38. Gordon Bell, Tony Hey, and Alex Szalay, "Beyond the Data Deluge," *Science* 323, no. 5919 (2009): 1297–98.

39. Vance Packard, *The Naked Society* (New York, D. McKay Co, 1964), 41, http://archive.org/details/nakedsociety00pack.

CHAPTER 7: INTELLIGENCE WITHOUT DATA

1. Claude Shannon to Irene Angus, 8 Aug. 1952, Shannon Papers, box 1, quoted in R. Kline, "Cybernetics, Automata Studies, and the Dartmouth Conference on Artificial Intelligence," *IEEE Annals of the History of Computing* 33, no. 4 (April 2011): 8, https://doi.org/10.1109/MAHC.2010 .44; For Shannon's aspirational survey of efforts in the early 1950s, see Claude E. Shannon, "Computers and Automata," *Proceedings of the IRE*

41, no. 10 (October 1953): 1234–41, https://doi.org/10.1109/JRPROC.1953.274273.

2. Much of the history of AI has been told in influential popular books. For the more academic history of AI, key major studies include Margaret A. Boden, *Mind as Machine: A History of Cognitive Science* (New York: Oxford University Press, 2006); Nils J. Nilsson, *The Quest for Artificial Intelligence: A History of Ideas and Achievements* (Cambridge, UK: Cambridge University Press, 2010); Roberto Cordeschi, *The Discovery of the Artificial: Behavior, Mind and Machines Before and Beyond Cybernetics* (Dordrecht: Springer, 2002).

3. See the discussion in Stephanie Dick, "After Math: (Re)Configuring Minds, Proof, and Computing in the Postwar United States" (PhD diss., Harvard University, 2015), 2–3, https://dash.harvard.edu/handle/1/14226096.

4. Alan M. Turing, "Computing Machinery and Intelligence," *Mind* 59, no. 236 (1950): 447.

5. Turing, "Computing Machinery and Intelligence," 449.

6. Turing, "Computing Machinery and Intelligence," 449.

7. Lucy A. Suchman, *Human-Machine Reconfigurations: Plans and Situated Actions*, 2nd ed. (Cambridge and New York: Cambridge University Press, 2007), 226; see Stephanie Dick, "AfterMath: The Work of Proof in the Age of Human–Machine Collaboration," *Isis* 102, no. 3 (2011): 495n3, https://doi.org/10.1086/661623.

8. John McCarthy et al., "A Proposal for the Dartmouth Summer Research Project on Artificial Intelligence," August 31, 1955, 16, Rockefeller Archive Center, Rockefeller Foundation records, projects, RG 1.2, series 200.D, box 26, folder 219.

9. Alma Steingart, *Axiomatics: Mathematical Thought and High Modernism* (Chicago: University of Chicago Press, forthcoming).

10. Claude Lévi-Strauss, "The Mathematics of Man," *International Social Science Bulletin* 6, no. 4 (1954): 586. Discussed in Steingart, *Axiomatics*.

11. Lévi-Strauss, 585.

12. Bruce G. Buchanan and Edward Hance Shortliffe, *Rule-Based Expert Systems: The MYCIN Experiments of the Stanford Heuristic Programming Project* (Reading, MA: Addison-Wesley, 1984), 3, http://archive.org/details/rulebasedexperts00buch.

13. John McCarthy, in "The General Purpose Robot is a Mirage," Controversy Programme, BBC, August 20, 1973, available as The Lighthill Debate (1973)—Part 4 of 6, https://www.youtube.com/watch?v=pyU9pm1hmYs&t=266s, at 4:26.

14. McCarthy et al., "A Proposal for the Dartmouth Summer Research Project on Artificial Intelligence."

15. The gulf between Shannon and McCarthy is best documented in Kline, "Cybernetics, Automata Studies, and the Dartmouth Conference on Artificial Intelligence"; Jonathan Penn, "Inventing Intelligence: On the History of Complex Information Processing and Artificial Intelligence in the United States in the Mid-Twentieth Century" (Thesis, University of Cambridge, 2021), 123–24, https://doi.org/10.17863/CAM.63087.

16. Penn, "Inventing Intelligence," 134. In other computer fields, like Human-Computer Interaction, professional psychologists figured much more centrally. See Sam Schirvar, "Machinery for Managers: Secretaries,

Psychologists, and 'Human-Computer Interaction', 1973–1983" (under review).

17. Pamela McCorduck, *Machines Who Think: A Personal Inquiry into the History and Prospects of Artificial Intelligence*, 25th anniversary update (Natick, MA: A.K. Peters, 2004), 114.

18. Hunter Heyck, "Defining the Computer: Herbert Simon and the Bureaucratic Mind—Part 1," *IEEE Annals of the History of Computing* 30, no. 2 (April 2008): 42–51, https://doi.org/10.1109/MAHC.2008.18; Hunter Heyck, "Defining the Computer: Herbert Simon and the Bureaucratic Mind—Part 2," *IEEE Annals of the History of Computing* 30, no. 2 (April 2008): 52–63, https://doi.org/10.1109/MAHC.2008.19.

19. Allen Newell, J. C. Shaw, and Herbert A. Simon, "Elements of a Theory of Human Problem Solving," *Psychological Review* 65, no. 3 (1958): 153, https://doi.org/10.1037/h0048495.

20. Penn, "Inventing Intelligence," 45.

21. Jamie Cohen-Cole, "The Reflexivity of Cognitive Science: The Scientist as Model of Human Nature," *History of the Human Sciences* 18, no. 4 (November 1, 2005): 122, https://doi.org/10.1177/0952695105058473; for AI's relationship with cognitive science more generally, Cordeschi, *The Discovery of the Artificial*.

22. For heuristics, compare Ekaterina Babintseva, "From Pedagogical Computing to Knowledge-Engineering: The Origins and Applications of Lev Landa's Algo-Heuristic Theory," in *Abstractions and Embodiments: New Histories of Computing and Society*, ed. Stephanie Dick and Janet Abbate (Baltimore, MD: Johns Hopkins University Press, 2022).

23. For a nuanced read of tensions of automation and freedom in mathematically focused AI, see Stephanie Dick, "The Politics of Representation: Narratives of Automation in Twentieth Century American Mathematics," in *Narrative Science: Reasoning, Representing and Knowing since 1800*, ed. Mary S. Morgan, Kim M. Hajek, and Dominic J. Berry (Cambridge University Press, forthcoming).

24. Discussion of John McCarthy, "Programs with Common Sense," in *Mechanisation of Thought Processes; Proceedings of a Symposium Held at the National Physical Laboratory on 24th, 25th, 26th and 27th November 1958*, ed. National Physical Laboratory (Great Britain) (London: H.M. Stationery Office, 1961), 86–87, 88, http://archive.org/details/mechanisationoft01nati.

25. Notably, Alison Adam, *Artificial Knowing: Gender and the Thinking Machine* (New York: Routledge, 1998).

26. Marvin Minsky, "Steps toward Artificial Intelligence," in *Computers and Thought*, ed. Edward A. Feigenbaum and Julian Feldman (New York, McGraw-Hill, 1963), 428, http://archive.org/details/computersthought00feig. The paper appeared originally in 1961.

27. Margaret A. Boden, "GOFAI," in *The Cambridge Handbook of Artificial Intelligence*, ed. Keith Frankish and William M. Ramsey (Cambridge, UK: Cambridge University Press, 2014), 89, https://doi.org/10.1017/CBO978 1139046855.007.

28. Jon Agar, "What Difference Did Computers Make?," *Social Studies of Science* 36, no. 6 (December 1, 2006): 898, https://doi.org/10.1177/0306312 706073450.

29. Penn, "Inventing Intelligence."
30. Stephanie Dick, "Of Models and Machines: Implementing Bounded Rationality," *Isis* 106, no. 3 (2015): 630.
31. Sir James Lighthill, "Lighthill Report," Artificial Intelligence: A General Survey, June 1972, http://www.chilton-computing.org.uk/inf/literature /reports/lighthill_report/p001.htm. For the context of the report see Jon Agar, "What Is Science for? The Lighthill Report on Artificial Intelligence Reinterpreted," *The British Journal for the History of Science* 53, no. 3 (September 2020): 289–310, https://doi.org/10.1017/S0007087420000230.
32. David C. Brock, "Learning from Artificial Intelligence's Previous Awakenings: The History of Expert Systems," *AI Magazine* 39, no. 3 (September 28, 2018): 3–15, https://doi.org/10.1609/aimag.v39i3.2809; David Ribes et al., "The Logic of Domains," *Social Studies of Science* 49, no. 3 (June 1, 2019): 287–91, https://doi.org/10.1177/0306312719849709; Hallam Stevens, "The Business Machine in Biology—The Commercialization of AI in the Life Sciences," *IEEE Annals of the History of Computing* 44, no. 01 (January 1, 2022): 8–19, https://doi.org/10.1109/MAHC.2021.3104868.
33. E. A. Feigenbaum, B. G. Buchanan, and J. Lederberg, "On Generality and Problem Solving: A Case Study Using the DENDRAL Program," *Machine Intelligence*, no. 6 (1971): 187.
34. Marvin Minsky and Seymour Papert, "Progress Report on Artificial Intelligence," 1971, Artificial Intelligence Memo AIM-252, https://web.media .mit.edu/~minsky/papers/PR1971.html.
35. Ira Goldstein and Seymour Papert, "Artificial Intelligence, Language, and the Study of Knowledge," *Cognitive Science* 1, no. 1 (January 1, 1977): 85, https://doi.org/10.1016/S0364-0213(77)80006-2.
36. Joseph Adam November, *Biomedical Computing: Digitizing Life in the United States* (Baltimore: Johns Hopkins University Press, 2012), 259–68.
37. Buchanan and Shortliffe, *Rule-Based Expert Systems*, 16.
38. For the bottleneck, see Stephanie A. Dick, "Coded Conduct: Making MACSYMA Users and the Automation of Mathematics," *BJHS Themes* 5 (ed. 2020): 205–24, https://doi.org/10.1017/bjt.2020.10; Edward Feigenbaum, Oral History, interview by Nils Nilsson, 20, 27 2007, 62–63, http:// archive.computerhistory.org/resources/access/text/2013/05/102702002-05 -01-acc.pdf; D. E. Forsythe, "Engineering Knowledge: The Construction of Knowledge in Artificial Intelligence," *Social Studies of Science* 23, no. 3 (August 1, 1993): 445–77, https://doi.org/10.1177/0306312793023003002.
39. J. R. Quinlan, "Discovering Rules by Induction from Large Collections of Examples," in *Expert Systems in the Micro-Electronic Age*, ed. Donald Michie (Edinburgh: Edinburgh University Press, 1979), 168.
40. Donald Michie, "Expert Systems Interview," *Expert Systems* 2, no. 1 (1985): 22.
41. For the hidden successes of expert systems, see Stevens, "The Business Machine in Biology—The Commercialization of AI in the Life Sciences."
42. Jacob T. Schwartz, *The Limits of Artificial Intelligence* (New York: Courant Institute of Mathematical Sciences, New York University, 1986), 30, http:// archive.org/details/limitsofartifici00schw.
43. Comments by Y. Bar-Hillel on McCarthy, National Physical Laboratory (Great Britain), *Mechanisation of Thought Processes; Proceedings of a Symposium Held at the National Physical Laboratory on 24th, 25th, 26th and 27th*

November 1958 (London, H.M. Stationery off., 1961), 85, http://archive.org/details/mechanisationoft01nati.

44. For Newman, see B. Jack Copeland, "Max Newman—Mathematician, Code Breaker, Computer Pioneer," in *Colossus: The Secrets of Bletchley Park's Codebreaking Computers*, ed. B. Jack Copeland (Oxford, UK: Oxford University Press, 2006), 176–88.

45. E. A. Newman, "An Analysis of Non Mathematical Data Processing," in *Mechanisation of Thought Processes; Proceedings of a Symposium Held at the National Physical Laboratory on 24th, 25th, 26th and 27th November 1958*, ed. National Physical Laboratory (Great Britain) (London: H.M. Stationery Office, 1961), 866, http://archive.org/details/mechanisationoft02nati.

46. Newman, 875.

47. Richard O. Duda and Peter E Hart, *Pattern Classification and Scene Analysis* (New York: Wiley, 1973).

CHAPTER 8: VOLUME, VARIETY, AND VELOCITY

1. R. Blair Smith, Oral History by Robina Mapstone (Charles Babbage Institute, May 1980), 27, 29, http://conservancy.umn.edu/handle/11299/107637.

2. Martin Campbell-Kelly, *From Airline Reservations to Sonic the Hedgehog: A History of the Software Industry* (Cambridge, MA: MIT Press, 2003), 43, http://www.loc.gov/catdir/toc/fy035/2002075351.html. See also https://www.ibm.com/ibm/history/ibm100/us/en/icons/sabre/team/

3. R. W. Parker, "The SABRE System," *Datamation* 11 (September 1965): 49. See Campbell-Kelly, *From Airline Reservations to Sonic the Hedgehog*, 41–45.

4. Privacy Protection Study Commission, *Personal Privacy in an Information Society: The Report of the Privacy Protection Study Commission*. (Washington: The Commission: For sale by the Supt. of Docs., US Govt. Print. Off., 1977), 4.

5. Quotation from Thomas J. Misa, *Digital State: The Story of Minnesota's Computing Industry* (Minneapolis: University of Minnesota Press, 2013), 64.

6. James W. Cortada, *The Digital Flood: The Diffusion of Information Technology across the U.S., Europe, and Asia* (New York: Oxford University Press, 2012), 49.

7. Samuel S. Snyder, "Computer Advances Pioneered by Cryptologic Organizations," *Annals of the History of Computing* 2, no. 1 (1980): 60–70, at 65. "SOLO holds the distinction of being the first completely transistorized computer in the United States."

8. Eckert-Mauchly Computer Corporation (EMCC), "UNIVAC System Advertisement," 1948, 2, 5, https://www.computerhistory.org/revolution/early-computer-companies/5/103/447?position=0. See the discussion in Arthur L. Norberg, *Computers and Commerce: A Study of Technology and Management at Eckert-Mauchly Computer Company, Engineering Research Associates, and Remington Rand, 1946–1957* (Cambridge, MA: MIT Press, 2005), 185–86.

9. Norberg, *Computers and Commerce*, 191. For more on the unreliability of tape, see Thomas Haigh, "The Chromium-Plated Tabulator: Institutionalizing an Electronic Revolution, 1954–1958," *IEEE Annals of the History of Computing* 23, no. 4 (2001): 86, 88.

10. J. Abbate, *Recoding Gender: Women's Changing Participation in Computing* (Cambridge, MA: MIT Press, 2012), 37–38.

11. For these efforts, see James W. Cortada, "Commercial Applications of the

Digital Computer in American Corporations, 1945–1995," *IEEE Annals of the History of Computing* 18, no. 2 (1996): 18–29; Haigh, "The Chromium-Plated Tabulator."

12. J. M. Juran, quoted in Richard G. Canning, *Electronic Data Processing for Business and Industry* (New York: Wiley, 1956), 316, https://catalog .hathitrust.org/Record/001118357.

13. Xerox advertisement, *Datamation*, 11 (September 1965), p. 76.

14. Control Data Corporation advertisement, *Datamation*, 11 (September 1965), p. 87.

15. Quoted in Haigh, "The Chromium-Plated Tabulator," 97.

16. Paul Edwards, *A Vast Machine: Computer Models, Climate Data, and the Politics of Global Warming* (Cambridge, MA: MIT Press, 2010), 111.

17. Martha Poon, "Scorecards as Devices for Consumer Credit: The Case of Fair, Isaac & Company Incorporated," *The Sociological Review* 55, no. 2_ suppl (October 2007): 284–306, https://doi.org/10.1111/j.1467-954X.2007 .00740.x.

18. Josh Lauer, "Encoding the Consumer: The Computerization of Credit Reporting and Credit Scoring," in *Creditworthy* (New York: Columbia University Press, 2017), 183, https://doi.org/10.7312/laue16808-009. For the longer history of gauging credit, see Rowena Olegario, *A Culture of Credit: Embedding Trust and Transparency in American Business* (Cambridge, MA: Harvard University Press, 2006).

19. For the developments of credit cards and forms of gauging creditworthiness, see Louis Hyman, *Debtor Nation: The History of America in Red Ink* (Princeton, NJ: Princeton University Press, 2011), ch 7.

20. "Datamation: Editor's Readout: Big Brother," *Datamation* 11 (October 1965): 23.

21. Packard, *The Naked Society*, 41. Our work on privacy is deeply indebted to Sarah Elizabeth Igo, *The Known Citizen: A History of Privacy in Modern America* (Cambridge, MA: Harvard University Press, 2018). Comments from Professor Igo sharpened this chapter immensely.

22. Packard, *The Naked Society*, 41.

23. Stanton Wheeler, ed., *On Record: Files and Dossiers in American Life* (New Brunswick, NJ: Transaction Books, 1976), 19–20.

24. Arthur R. Miller, *The Assault on Privacy: Computers, Data Banks, and Dossiers* (New York: New American Library, 1972), 22.

25. United States Congress Senate Committee on Government Operations Ad Hoc Subcommittee on Privacy and Information, *Privacy: The Collection, Use, and Computerization of Personal Data: Joint Hearings Before the Ad Hoc Subcommittee on Privacy and Information Systems of the Committee on Government Operations and the Subcommittee on Constitutional Rights of the Committee on the Judiciary, United States Senate, Ninety-Third Congress, Second Session . . . June 18, 19, and 20, 1974* (Washington: US Govt. Print. Off, 1974), I:53.

26. Dan Bouk, "The National Data Center and the Rise of the Data Double," *Historical Studies in the Natural Sciences* 48, no. 5 (November 1, 2018): 627–36, https://doi.org/10.1525/hsns.2018.48.5.627; Igo, *The Known Citizen*; Priscilla M. Regan, *Legislating Privacy: Technology, Social Values, and Public Policy* (Chapel Hill: University of North Carolina Press, 1995), 71–73.

27. United States Congress Senate Committee on Government Operations Ad

Hoc Subcommittee on Privacy and Information, *Senate Ad Hoc Committee Privacy*, II:1739.

28. United States Congress Senate Committee on Government Operations Ad Hoc Subcommittee on Privacy and Information, II:1741.

29. Testimony of Alan Westin, United States Congress Senate Committee on Government Operations Ad Hoc Subcommittee on Privacy and Information, I:77-78.

30. Quoted in United States Congress Senate Committee on Government Operations Ad Hoc Subcommittee on Privacy and Information, *Senate Ad Hoc Committee Privacy*, I:651.

31. Statement of the National Bank Americard, United States Congress Senate Committee on Government Operations Ad Hoc Subcommittee on Privacy and Information, *Senate Ad Hoc Committee Privacy*. I:606.

32. United States Congress Senate Committee on Government Operations Ad Hoc Subcommittee on Privacy and Information, *Senate Ad Hoc Committee Privacy*, I:658.

33. Igo, *The Known Citizen*, 362–63.

34. Subcommittee on Courts, Civil Liberties, and the Administration of Justice of the Committee on the Judiciary, *1984, Civil Liberties and the National Security State: Hearings before the Subcommittee on Courts, Civil Liberties, and the Administration of Justice of the Committee on the Judiciary. House of Representatives, Ninety-Eighth Congress, First and Second Sessions, on 1984: Civil Liberties and the National Security State, November 2, 3, 1983 and January 24, April 5, and September 26, 1984.* (Washington, DC: US G.P.O., 1986), 267–79.

35. Subcommittee on Courts, Civil Liberties, and the Administration of Justice of the Committee on the Judiciary, *1984, Civil Liberties and the National Security State*, 294–95.

36. Office of Technology Assessment, *Electronic Record Systems and Individual Privacy*, 11.

37. Office of Technology Assessment, *Electronic Record Systems and Individual Privacy*, 39, 40.

38. United States and Privacy Protection Study Commission, *Personal Privacy in an Information Society*, 533 (original all in italics).

39. See, e.g., Eubanks, *Automating Inequality*.

40. Bobbie Johnson and Las Vegas, "Privacy No Longer a Social Norm, Says Facebook Founder," *The Guardian*, January 11, 2010, sec. Technology, https://www.theguardian.com/technology/2010/jan/11/facebook-privacy.

41. W. Lee Burge, "The Free Flow of Information: Key to Our Credit Economy" in United States Congress Senate Committee on Government Operations Ad Hoc Subcommittee on Privacy and Information, *Privacy: The Collection, Use, and Computerization of Personal Data: Joint Hearings Before the Ad Hoc Subcommittee on Privacy and Information Systems of the Committee on Government Operations and the Subcommittee on Constitutional Rights of the Committee on the Judiciary, United States Senate, Ninety-Third Congress, Second Session . . . June 18, 19, and 20, 1974* (Washington, DC: US Govt. Print. Off, 1974), I:650.

42. Jennifer Barrett Glasgow, "Acxiom, Letter to Representative Edward J. Markey," August 15, 2012, http://markey.house.gov/sites/markey.house.gov/files/documents/Acxiom.pdf.

43. New and Castro, "How Policymakers Can Foster Algorithmic Accountability," 2.
44. Paul Baran, "Legislation, Privacy and EDUCOM," *EDUCOM: Bulletin of the Interuniversity Communications Council*, December 3, 1968, 3.
45. Paul Baran, "On the Future Computer Era: Modification of the American Character and the Role of the Engineer, or, A Little Caution in the Haste to Number," RAND Paper (RAND Corporation, 1968), 14, https://www.rand.org/pubs/papers/P3780.html.
46. Robert Nozick, *Anarchy, State, and Utopia* (New York: Basic Books, 1974), 32–33.
47. Daniel T. Rodgers, *Age of Fracture* (Cambridge, MA: Harvard University Press, 2011), 190.
48. Milton Friedman, "The Social Responsibility of Business Is to Increase Its Profits (1970)," in *Corporate Ethics and Corporate Governance*, ed. Walther Ch Zimmerli, Markus Holzinger, and Klaus Richter (Berlin, Heidelberg: Springer, 2007), 178, https://doi.org/10.1007/978-3-540-70818-6_14.
49. See Jodi L. Short, "The Paranoid Style in Regulatory Reform," *Hastings Law Journal* 63 (2012): 633–94; Julie E. Cohen, *Between Truth and Power: The Legal Constructions of Informational Capitalism* (New York: Oxford University Press, 2019), 189; Amy Kapczynski, "The Law of Informational Capitalism," *The Yale Law Journal*, 2020, 1491.
50. Priscilla M. Regan, *Legislating Privacy: Technology, Social Values, and Public Policy* (Chapel Hill: University of North Carolina Press, 1995), 4.
51. Oscar H. Gandy, *The Panoptic Sort: A Political Economy of Personal Information*, Critical Studies in Communication and in the Cultural Industries (Boulder, CO: Westview, 1993).
52. Matthew Crain, *Profit Over Privacy: How Surveillance Advertising Conquered the Internet* (Minneapolis: University of Minnesota Press, 2021), 20.
53. Meg Leta Jones, "Cookies: A Legacy of Controversy," *Internet Histories* 4, no. 1 (January 2, 2020): 87–104, https://doi.org/10.1080/24701475.2020.1725852.
54. Quoted in Joshua Quittner, "The Merry Pranksters Go to Washington," *Wired*, accessed May 14, 2021, https://www.wired.com/1994/06/eff/; discussed in Fred Turner, *From Counterculture to Cyberculture: Stewart Brand, the Whole Earth Network, and the Rise of Digital Utopianism* (Chicago: University of Chicago Press, 2006), 219.
55. Turner, *From Counterculture to Cyberculture*, 261.
56. Paul Sabin, *Public Citizens: The Attack on Big Government and the Remaking of American Liberalism* (New York: Norton, 2021).
57. See Kapczynski, "The Law of Informational Capitalism," 1493–99; drawing on Cohen, *Between Truth and Power*.
58. Compare the dramatic shift from a broad understanding of professional responsibility toward a narrower civil rights focus, discussed in Megan Finn and Quinn DuPont, "From Closed World Discourse to Digital Utopianism: The Changing Face of Responsible Computing at Computer Professionals for Social Responsibility (1981–1992)," *Internet Histories* 4, no. 1 (January 2, 2020): 6–31, https://doi.org/10.1080/24701475.2020.1725851.
59. Quotation from secret decision with redacted name and date, p. 63, quoted in Amended Memorandum Opinion at 8–9.
60. Amended Memorandum Opinion at 8.

61. Felten, "Declaration of Professor Edward W. Felten in ACLU et al. *v*. James R. Clapper et al.," 8.
62. Anonymous, "Lessons Learned. Interview with [Redacted]," 1.
63. For challenges for consent, see, for example, Frank Pasquale, "Licensure as Data Governance," Knight First Amendment Institute at Columbia University, September 28, 2021, https://knightcolumbia.org/content/licensure-as -data-governance.
64. R Allen Wilkinson et al., *The First Census Optical Character Recognition System Conference*, NIST IR 4912 (Gaithersburg, MD: National Institute of Standards and Technology, 1992), 1, https://doi.org/10.6028/NIST.IR .4912.
65. Wilkinson et al., 4.
66. Wilkinson et al., 2.

CHAPTER 9: MACHINES, LEARNING

1. Pat Langley, "The Changing Science of Machine Learning," *Machine Learning* 82, no. 3 (March 2011): 277, https://doi.org/10.1007/s10994-011-5242 -y. For histories of prediction and machine learning, see Adrian Mackenzie, "The Production of Prediction: What Does Machine Learning Want?," *European Journal of Cultural Studies* 18, no. 4–5 (2015): 429–45; Ann Johnson, "Rational and Empirical Cultures of Prediction," in *Mathematics as a Tool*, ed. Johannes Lenhard and Martin Carrier, vol. 327 (Cham, Switzerland: Springer International Publishing, 2017), 23–35, https://doi.org/10 .1007/978-3-319-54469-4_2; Adrian Mackenzie, *Machine Learners: Archaeology of a Data Practice* (Cambridge, MA: MIT Press, 2018); Aaron Plasek, "On the Cruelty of Really Writing a History of Machine Learning," *IEEE Annals of the History of Computing* 38, no. 4 (December 2016): 6–8, https:// doi.org/10.1109/MAHC.2016.43; Aaron Mendon-Plasek, "Mechanized Significance and Machine Learning: Why It Became Thinkable and Preferable to Teach Machines to Judge the World," in *The Cultural Life of Machine Learning: An Incursion into Critical AI Studies*, ed. Jonathan Roberge and Michael Castelle (Cham, Switzerland: Springer International Publishing, 2021), 31–78, https://doi.org/10.1007/978-3-030-56286-1_2; Cosma Rohilla Shalizi, "The Formation of the Statistical Learning Paradigm and the Field of Machine Learning, c. 1985-2000" (2020), manuscript in preparation, available on request from the author.
2. P. Langley and J. G. Carbonell, "Approaches to Machine Learning," *Journal of the American Society for Information Science* 35, no. 5 (September 1, 1984): 306–16, at 306.
3. Rosenblatt, "The Perceptron: A Perceiving and Recognizing Automaton (Project PARA)," 1, https://blogs.umass.edu/brain-wars/files/2016/03 /rosenblatt-1957.pdf.
4. Jonathan Penn, "Inventing Intelligence: On the History of Complex Information Processing and Artificial Intelligence in the United States in the Mid-Twentieth Century" (Thesis, University of Cambridge, 2021), 96–98, https://doi.org/10.17863/CAM.63087. For Rosenblatt and contemporaneous economic projects, check out Orit Halpern, "The Future Will Not Be Calculated: Neural Nets, Neoliberalism, and Reactionary Politics," *Critical Inquiry* 48, no. 2 (January 1, 2022): 334–59, https://doi .org/10.1086/717313.

5. "New Navy Device Learns by Doing; Psychologist Shows Embryo of Computer Designed to Read and Grow Wiser," *New York Times*, July 8, 1958, http://timesmachine.nytimes.com/timesmachine/1958/07/08/83417341.html.

6. Herbert A. Simon, "Why Should Machines Learn?," in *Machine Learning: An Artificial Intelligence Approach*, ed. Ryszard S. Michalski, Jaime G. Carbonell, and Tom M. Mitchell, Symbolic Computation (Berlin, Heidelberg: Springer, 1983), 32, https://doi.org/10.1007/978-3-662-12405-5_2.

7. For an overview of these efforts, with special attention to activities around Stanford, see Nils J. Nilsson, *The Quest for Artificial Intelligence: A History of Ideas and Achievements* (Cambridge, UK: Cambridge University Press, 2010), ch. 4.

8. For the importance of this "gray area" of research between government and academia, see Joy Rohde, *Armed with Expertise: The Militarization of American Social Research during the Cold War* (Ithaca, NY: Cornell University Press, 2013).

9. Laveen N. Kanal, "Preface," in *Pattern Recognition*, ed. Laveen N. Kanal (Washington, DC: Thompson Book Co, 1968), xi.

10. G. Nagy, "State of the Art in Pattern Recognition," *Proceedings of the IEEE* 56, no. 5 (May 1968): 836–63, https://doi.org/10.1109/PROC.1968.6414.

11. Xiaochang Li, " 'There's No Data Like More Data': Automatic Speech Recognition and the Making of Algorithmic Culture," in *Osiris, "Beyond Craft and Code,"* ed. James Evans and Adrian Johns, forthcoming, captures this crucial point.

12. Mendon-Plasek, "Mechanized Significance and Machine Learning," 2–3. Michael Castelle is finishing a major history of the loss function.

13. J. McCarthy et al., "A Proposal for the Dartmouth Summer Research Project on Artificial Intelligence," August 31, 1955, Rockefeller Archive Center, Rockefeller Foundation records, projects, RG 1.2, series 200.D, box 26, folder 219.

14. Sir James Lighthill, "Lighthill Report," Artificial Intelligence: A General Survey, June 1972, §3, http://www.chilton-computing.org.uk/inf/literature/reports/lighthill_report/p001.htm.

15. Rodney A. Brooks, "Intelligence without Representation," *Artificial Intelligence* 47, no. 1–3 (1991): 140.

16. For the hidden success of expert systems, see Hallam Stevens, "The Business Machine in Biology—The Commercialization of AI in the Life Sciences," *IEEE Annals of the History of Computing* 44, no. 01 (January 1, 2022): 8–19, https://doi.org/10.1109/MAHC.2021.3104868.

17. Keki B. Irani et al., "Applying Machine Learning to Semiconductor Manufacturing," *IEEE Expert* 8, no. 1 (1993): 41.

18. Alain Desrosières, *Prouver et gouverner: une analyse politique des statistiques publiques* (Paris: Découverte, 2014), ch. 9.

19. Hayashi Chikiō and M. Takahashi, "Kagakusi to Kagakusha: Hayashi Chikiosi Kōkai Intabyu," *Kōdō Keiryōgaku* 31, no. 2 (2004): 107–24; quoted and translated in Joonwoo Son, "Data Science in Japan" (Unpublished MS, Columbia University, Sociology, May 2016).

20. Vladimir Naumovich Vapnik, *Estimation of Dependences Based on Empirical Data (1982); Empirical Inference Science: Afterword of 2006*, 2nd ed. (New York: Springer, 2006), 415.

21. See "History of the Institute | ИПУ РАН," accessed July 7, 2017, http://www.ipu.ru/en/node/12744.

22. Vapnik has often been highly critical of the mathematical and theoretical insufficiencies of most other data analysis. For an appreciation, see Léon Bottou, "In Hindsight: Doklady Akademii Nauk SSSR, 181 (4), 1968," in *Empirical Inference* (Springer, 2013), 3–5, http://link.springer.com/chapter /10.1007/978-3-642-41136-6_1.

23. Xiaochang Li, "Divination Engines: A Media History of Text Prediction" (NYU, 2017); Xiaochang Li and Mara Mills, "Vocal Features: From Voice Identification to Speech Recognition by Machine," *Technology and Culture* 60, no. 2 (June 18, 2019): S129–60, https://doi.org/10.1353/tech.2019.0066.

24. For compelling autobiographical accounts, see Terrence J. Sejnowski, *The Deep Learning Revolution* (Cambridge, MA: MIT Press, 2018); Yann LeCun, *Quand la machine apprend: La Révolution des neurones artificiels et de l'apprentissage profond* (Paris: Odile Jacob, 2019); for a fine journalistic account, see John Markoff, *Machines of Loving Grace: The Quest for Common Ground between Humans and Robots*, 2016; and the excellent account Dominique Cardon, Jean-Philippe Cointet, and Antoine Mazières, "La revanche des neurones: L'invention des machines inductives et la controverse de l'intelligence artificielle," *Réseaux* n° 211, no. 5 (2018): 173, https://doi.org/10.3917/res.211.0173.

25. While versions of the approach were published earlier, the key transformative studies include David E. Rumelhart and James L. McClelland, "Learning Internal Representations by Error Propagation," in *Parallel Distributed Processing: Explorations in the Microstructure of Cognition: Foundations* (Cambridge, MA: MIT Press, 1987), 318–62, http://ieeexplore .ieee.org/document/6302929; David E. Rumelhart, Geoffrey E. Hinton, and Ronald J. Williams, "Learning Representations by Back-Propagating Errors," *Nature* 323, no. 6088 (October 1986): 533–36, https://doi.org/10 .1038/323533a0; Y. LeCun et al., "Backpropagation Applied to Handwritten Zip Code Recognition," *Neural Computation* 1, no. 4 (December 1989): 541–51, https://doi.org/10.1162/neco.1989.1.4.541; P. J. Werbos, "Backpropagation through Time: What It Does and How to Do It," *Proceedings of the IEEE* 78, no. 10 (October 1990): 1550–60, https://doi.org/10.1109/5.58337.

26. For many memorable stories of this period, see LeCun, *Quand la machine apprend*.

27. From an anonymous interviewee in Cardon, Cointet, and Mazières, "La revanche des neurones," 21.

28. Leo Breiman and Nong Shang, "Born Again Trees," n.d., https://www.stat .berkeley.edu/~breiman/BAtrees.pdf.

29. For the revival of neural nets, see Yann LeCun, Yoshua Bengio, and Geoffrey Hinton, "Deep Learning," *Nature* 521, no. 7553 (May 27, 2015): 436–44, https://doi.org/10.1038/nature14539.

30. Alex Krizhevsky, Ilya Sutskever, and Geoffrey E. Hinton, "ImageNet Classification with Deep Convolutional Neural Networks," in *Advances in Neural Information Processing Systems*, vol. 25 (Curran Associates, Inc., 2012), https://papers.nips.cc/paper/2012/hash/c399862d3b9d6 b76c8436e924a68c45b-Abstract.html.

31. Olga Russakovsky et al., "ImageNet Large Scale Visual Recognition Chal-

lenge," *International Journal of Computer Vision* 115, no. 3 (December 1, 2015): 211–52, https://doi.org/10.1007/s11263-015-0816-y.

32. Fei-Fei Li, "Crowdsourcing, Benchmarking & Other Cool Things," https://web.archive.org/web/20121110041643/http://www.image-net.org/papers/ImageNet_2010.pdf; Hao Su, Jia Deng, and Li Fei-Fei, "Crowdsourcing Annotations for Visual Object Detection," n.d., 7.

33. For the deep problems and controversies over this data set, see Kate Crawford and Trevor Paglen, "Excavating AI," September 19, 2019, https://excavating.ai.

34. Cardon, Cointet, and Mazières, "La revanche des neurones."

35. The precise scope of the environmental toll is much debated. Scholars looking at the environmental and infrastructural costs of this massive computing include Mél Hogan, "Data Flows and Water Woes: The Utah Data Center," *Big Data & Society* 2, no. 2 (December 1, 2015): 2053951715592429, https://doi.org/10.1177/2053951715592429; Nathan Ensmenger, "The Environmental History of Computing," *Technology and Culture* 59, no. 4 (2018): S7–33, https://doi.org/10.1353/tech.2018.0148; Kate Crawford, *Atlas of AI: Power, Politics, and the Planetary Costs of Artificial Intelligence* (New Haven, CT: Yale University Press, 2021), https://doi.org/10.12987/9780300252392, ch. 1.

36. Meredith Whittaker, "The Steep Cost of Capture," *Interactions* 28, no. 6 (November 2021): 52, https://doi.org/10.1145/3488666.

37. M. I. Jordan and T. M. Mitchell, "Machine Learning: Trends, Perspectives, and Prospects," *Science* 349, no. 6245 (July 17, 2015): 255–60, https://doi.org/10.1126/science.aaa8415.

38. Jordan and Mitchell.

39. Langley, "The Changing Science of Machine Learning," 278.

40. See Jenna Burrell, "How the Machine 'Thinks': Understanding Opacity in Machine Learning Algorithms," *Big Data & Society* 3, no. 1 (January 5, 2016): 4–5, https://doi.org/10.1177/2053951715622512.

41. "Netflix Prize: Review Rules," February 2, 2007, https://web.archive.org/web/20070202023620/http://www.netflixprize.com:80/rules.

42. Quoted in Steve Lohr, "A $1 Million Research Bargain for Netflix, and Maybe a Model for Others," *New York Times*, September 22, 2009, sec. Technology, https://www.nytimes.com/2009/09/22/technology/internet/22netflix.html.

43. David Donoho, "50 Years of Data Science," *Journal of Computational and Graphical Statistics* 26, no. 4 (October 2, 2017): 752, https://doi.org/10.1080/10618600.2017.1384734.

44. Donoho, 752–53.

45. Quoted in Lohr, "A $1 Million Research Bargain for Netflix, and Maybe a Model for Others."

CHAPTER 10: THE SCIENCE OF DATA

1. Allen Ginsberg, "Howl," text/html, Poetry Foundation (Poetry Foundation, August 12, 2021), https://www.poetryfoundation.org/, https://www.poetryfoundation.org/poems/49303/howl.

2. Ashlee Vance, "In Ads We Trust," *Bloomberg Businessweek*, no. 4521 (May 8, 2017): 6.

3. Chris Anderson, "The Long Tail," *Wired*, October 2004, https://www.wired.com/2004/10/tail/.
4. Gregory Zuckerman, *The Man Who Solved the Market: How Jim Simons Launched the Quant Revolution* (New York: Penguin, 2019). For deep relationship between the development of speech recognition technologies and financial markets, Xiaochang Li, "Divination Engines: A Media History of Text Prediction" (PhD Thesis, NYU, 2017).
5. Ognjenka Goga Vukmirovic and Shirley M. Tilghman, "Exploring Genome Space," *Nature* 405, no. 6788 (June 2000): 820, https://doi.org/10.1038/35015690. Generally see Hallam Stevens, *Life Out of Sequence: A Data-Driven History of Bioinformatics* (Chicago: University of Chicago Press, 2013); Sabina Leonelli, *Data-Centric Biology: A Philosophical Study* (Chicago: University of Chicago Press, 2016), 18.
6. Cathy O'Neil, "Data Science: Tools vs. Craft," *Mathbabe* (blog), October 4, 2011, https://mathbabe.org/2011/10/04/data-science-tools-vs-craft/.
7. Cosma Rohilla Shalizi, "New 'Data Scientist' Is But Old 'Statistician' Writ Large," December 4, 2011, https://web.archive.org/web/20111204161344/http://cscs.umich.edu/~crshalizi/weblog/805.html.
8. Solomon Kullback to Tukey, 13.3.1959, American Philosophical Society [APS] Tukey Papers, Ms 117, Series I: US: NSA.
9. Howard Barlow for Solomon Kullback to John Tukey, 6.4.1959, APS Tukey Papers, Ms 117, Series I: US: NSA.
10. Solomon Kullback to Tukey, 13.3.1959, APS Tukey Papers, Ms 117, Series I: US: NSA.
11. John W. Tukey, "The Future of Data Analysis," *The Annals of Mathematical Statistics* 33, no. 1 (1962): 6, italics his.
12. Luisa T. Fernholz et al., "A Conversation with John W. Tukey and Elizabeth Tukey," *Statistical Science* 15, 2000, 80–81.
13. John W. Tukey, *Exploratory Data Analysis* (Addison Wesley, 1977), 2–3; for Tukey's project see Alexander Campolo, "Steering by Sight: Data, Visualization, and the Birth of an Informational Worldview" (PhD diss., New York University, 2019), 186–88.
14. Tukey, *Exploratory Data Analysis*, 56.
15. John M. Chambers, "Greater or Lesser Statistics: A Choice for Future Research," *Statistics and Computing* 3, no. 4 (1993): 182.
16. Chambers, 182.
17. Chambers, 183.
18. For examples of influential articles near the end of the first decade of this century arguing that the abundance of data would usher in new ways of understanding science, see Chris Anderson, "The End of Theory: The Data Deluge Makes the Scientific Method Obsolete," *Wired*, 2008, http://archive.wired.com/science/discoveries/magazine/16-07/pb_theory; Tony Hey, Stewart Tansley, and Kristin Tolle, *The Fourth Paradigm: Data-Intensive Scientific Discovery*, *The Fourth Paradigm: Data-Intensive Scientific Discovery* (Microsoft Research, 2009), https://www.microsoft.com/en-us/research/publication/fourth-paradigm-data-intensive-scientific-discovery/.
19. "John M. Chambers," https://awards.acm.org/award_winners/chambers_6640862.
20. William S. Cleveland, "Data Science: An Action Plan for Expanding the

Technical Areas of the Field of Statistics," *International Statistical Review / Revue Internationale de Statistique* 69, no. 1 (April 2001): 23, https://doi.org/10.2307/1403527.

21. The central conference in the field, first convened in 1975, was the International Conference on Very Large Data Bases.

22. Usama Fayyad, "Mining Databases: Towards Algorithms for Knowledge Discovery," *Bulletin of the Technical Committee on Data Engineering* 21, no. 1 (1998): 48.

23. See Usama M. Fayyad, Gregory Piatetsky-Shapiro, and Padhraic Smyth, "From Data Mining to Knowledge Discovery: An Overview," in *Advances in Knowledge Discovery and Data Mining* (Menlo Park, CA: AAAI/MIT Press, 1996), 1–34.

24. Matthew L. Jones, "Querying the Archive: Database Mining from Apriori to Page-Rank," in *Science in the Archives: Pasts, Presents, Futures*, ed. Lorraine Daston (Chicago: University of Chicago Press, 2016), 311–28.

25. Shawn Thelen, Sandra Mottner, and Barry Berman, "Data Mining: On the Trail to Marketing Gold," *Business Horizons* 47 (2004): 26, https://doi.org/10.1016/j.bushor.2004.09.005.

26. Patrick O. Brown and David Botstein, "Exploring the New World of the Genome with DNA Microarrays," *Nature Genetics* 21 (January 1, 1999): 26, https://doi.org/10.1038/4462.

27. The list of seminars as of the end of 1998 are available at http://web.archive.org/web/19990116232602/http://www.almaden.ibm.com/cs/quest/seminars.html and http://web.archive.org/web/19980210042739/http://www.almaden.ibm.com/cs/quest/seminars-hist.html.

28. The webpage for MIDAS is preserved at http://infolab.stanford.edu/midas/; a listserv of the data mining group can be found on Yahoo e-groups. See Jeffrey D. Ullman, "The MIDAS Data-Mining Project at Stanford," in *Database Engineering and Applications, 1999. IDEAS'99. International Symposium Proceedings*, 1999, 460–64, http://ieeexplore.ieee.org/xpls/abs_all.jsp?arnumber=787298.

29. A printed version of this material appeared as Sergey Brin, Rajeev Motwani, and Terry Winograd, "What Can You Do with a Web in Your Pocket," *Data Engineering Bulletin* 21 (1998): 37–47.

30. http://infolab.stanford.edu/midas/.

31. Thomas Haigh, "The Web's Missing Links: Search Engines and Portals," in *The Internet and American Business*, ed. William Aspray and Paul Ceruzzi (Cambridge, MA: MIT Press, 2008), 160–61. Sergey Brin and Lawrence Page, "The Anatomy of a Large-Scale Hypertextual Web Search Engine," in *Seventh International World-Wide Web Conference (WWW 1998)*, 1998, http://ilpubs.stanford.edu:8090/361/, §3.1. ". . . most of the research on information retrieval systems is on small well controlled homogeneous collections such as collections of scientific papers or news stories on a related topic."

32. Sergey Brin and Lawrence Page, "Dynamic Data Mining: Exploring Large Rule Spaces by Sampling," Technical Report (Stanford InfoLab, November 1999), 2, http://ilpubs.stanford.edu:8090/424/.

33. Brin and Page, "The Anatomy of a Large-Scale Hypertextual Web Search Engine," §4.2.

34. Brin and Page, §4.2.

35. *Statistical Analysis of Massive Data Streams* (National Research Council of the National Academies, 2004), 8–9.
36. Alon Halevy, Fernando Pereira, and Peter Norvig, "The Unreasonable Effectiveness of Data," *Intelligent Systems, IEEE* 24, no. 2 (April 2009): 8–12, https://doi.org/10.1109/MIS.2009.36.
37. Redacted, "Confronting the Intelligence Future (U) An Interview with William P. Crowell, NSA's Deputy Director (U)," *Cryptolog* 22, no. 2 (1996): 1–5.
38. Paul Burkhardt and Chris Waring, "An NSA Big Graph Experiment."
39. NSA Job ID: 1034503
40. Redacted, "NSA Culture, 1980s to the 21st Century—a SID Perspective," *Cryptological Quarterly* 30, no. 4 (2011): 84. Bulleted points are rendered as continuous prose.
41. Catherine D'Ignazio and Lauren F. Klein, *Data Feminism* (Cambridge, MA: MIT Press, 2020), 173.
42. Antonio A. Casilli, *En attendant les robots: enquête sur le travail du clic* (Paris: Éditions du Seuil, 2019), 14.
43. Sarah T. Roberts, *Behind the Screen: Content Moderation in the Shadows of Social Media* (New Haven, CT: Yale University Press, 2019).
44. Mary L. Gray and Siddharth Suri, *Ghost Work: How to Stop Silicon Valley from Building a New Global Underclass* (Boston: Houghton Mifflin Harcourt, 2019).
45. Casilli, *En attendant les robots*, 17.
46. Lilly Irani, "The Cultural Work of Microwork," *New Media & Society* 17, no. 5 (May 2015): 723, https://doi.org/10.1177/1461444813511926.
47. Lilly Irani, "Justice for 'Data Janitors,' " *Public Books* (blog), January 15, 2015, http://www.publicbooks.org/justice-for-data-janitors/.
48. Bin Yu, "Institute of Mathematical Statistics | IMS Presidential Address: Let Us Own Data Science," July 2014, https://imstat.org/2014/10/01/ims-presidential-address-let-us-own-data-science/.
49. Richard Olshen and Leo Breiman, "A Conversation with Leo Breiman," *Statistical Science*, 2001, 196.
50. Olshen and Breiman, 188.
51. Leo Breiman, "[A Report on the Future of Statistics]: Comment," *Statistical Science* 19, no. 3 (2004): 411–411.
52. Leo Breiman, "Statistical Modeling: The Two Cultures," *Statistical Science* 16, no. 3 (2001): 201.
53. Chambers, "Greater or Lesser Statistics," 182.
54. David Madigan and Werner Stuetzle, "[A Report on the Future of Statistics]: Comment," *Statistical Science* 19, no. 3 (2004): 408.
55. See the much-cited McKinsey report Nicolaus Henke and Jacques Bughin, "The Age of Analytics: Competing in a Data-Driven World" (McKinsey Global Institute, December 2016).
56. Gina Neff et al., "Critique and Contribute: A Practice-Based Framework for Improving Critical Data Studies and Data Science," *Big Data* 5, no. 2 (June 2017): 85–97, https://doi.org/10.1089/big.2016.0050.
57. Jennifer Bryan and Hadley Wickham, "Data Science: A Three Ring Circus or a Big Tent?," *Journal of Computational and Graphical Statistics* 26, no. 4 (October 2, 2017): 784–85, https://doi.org/10.1080/10618600.2017.1389743.
58. "Are Data Scientists at Facebook Really Data Analysts," 25.8.2017, https://

www.reddit.com/r/datascience/comments/6vv7u2/are_data_scientists_at
_facebook_really_data/

59. Ryan Tibshirani, "Delphi's COVIDcast Project: Lessons from Building a Digital Ecosystem for Tracking and Forecasting the Pandemic," https:// docs.google.com/presentation/d/1t_T8BRIkvC5CDOgE4_1PekPw-SThN2 nMJTdieYgdnr4.

60. Blaise Aguera y Arcas, Margaret Mitchell, and Alexander Todorov, "Physiognomy's New Clothes," *Medium* (blog), May 20, 2017, https://medium .com/@blaisea/physiognomys-new-clothes-f2d4b59fdd6a.

61. Luke Stark and Jevan Hutson, "Physiognomic Artificial Intelligence," *Fordham Intellectual Property, Media & Entertainment Law Journal*, no. forthcoming, https://doi.org/10.2139/ssrn.3927300, p. 5 (internal citations removed).

62. D'Ignazio and Klein, *Data Feminism*, esp. ch. 2.

63. Brin to listserv 10.11.97. The webpage for the Stanford group MIDAS is preserved at http://infolab.stanford.edu/midas/; a partial archive of the listserv of the data mining group was available on Yahoo e-groups before it was deleted; Jones has a partial copy. Brin planned to copy old messages but failed to do so.

CHAPTER 11: THE BATTLE FOR DATA ETHICS

1. AI Now Institute, *Austerity, Inequality, and Automation | AI Now 2018 Symposium*, 2018, https://www.youtube.com/watch?v=gI1KxTrPDLo.

2. Joy Buolamwini and Timnit Gebru, "Gender Shades: Intersectional Accuracy Disparities in Commercial Gender Classification," in *Proceedings of the 1st Conference on Fairness, Accountability and Transparency* (Conference on Fairness, Accountability and Transparency, PMLR, 2018), 77–91, https:// proceedings.mlr.press/v81/buolamwini18a.html. See also accompanying website for the project, http://gendershades.org/.

3. Margaret Mitchell et al., "Model Cards for Model Reporting," in *Proceedings of the Conference on Fairness, Accountability, and Transparency*, FAT* '19 (New York: Association for Computing Machinery, 2019), 220–29, https://doi.org/10.1145/3287560.3287596.

4. Tom Simonite, "Google Offers to Help Others With the Tricky Ethics of AI," *Wired*, accessed August 24, 2021, https://www.wired.com/story/google -help-others-tricky-ethics-ai/.

5. Cade Metz and Daisuke Wakabayashi, "Google Researcher Says She Was Fired Over Paper Highlighting Bias in A.I.," *New York Times*, December 3, 2020, sec. Technology, https://www.nytimes.com/2020/12/03/technology /google-researcher-timnit-gebru.html.

6. Adam D. I. Kramer, Jamie E. Guillory, and Jeffrey T. Hancock, "Experimental Evidence of Massive-Scale Emotional Contagion through Social Networks," *Proceedings of the National Academy of Sciences* 111, no. 24 (June 17, 2014): 8788–90, https://doi.org/10.1073/pnas.1320040111.

7. Alex Hern, "Facebook Deliberately Made People Sad. This Ought to Be the Final Straw," *The Guardian*, June 30, 2014, sec. Opinion, https://www .theguardian.com/commentisfree/2014/jun/30/facebook-sad-manipulating -emotions-socially-responsible-company.

8. Matt Murray, "Users Angered at Facebook Emotion-Manipulation Study," TODAY.com, June 30, 2014, http://www.today.com/health/users-angered -facebook-emotion-manipulation-study-1D79863049.Murray.

9. M. J. Salganik, *Bit by Bit: Social Research in the Digital Age* (Princeton, NJ: Princeton University Press, 2017), 282.

10. Chris Chambers, "Facebook Fiasco: Was Cornell's Study of 'Emotional Contagion' an Ethics Breach?," *The Guardian*, July 1, 2014, https://www .theguardian.com/science/head-quarters/2014/jul/01/facebook-cornell -study-emotional-contagion-ethics-breach.

11. Allan M. Brandt, "Racism and Research: The Case of the Tuskegee Syphilis Study," *Hastings Center Report* 8, no. 6 (1978): 21–29, https://doi.org/10 .2307/3561468.

12. R. A. Vonderlehr et al., "Untreated Syphilis in the Male Negro: A Comparative Study of Treated and Untreated Cases," *Journal of the American Medical Association* 107, no. 11 (September 12, 1936): 856–60, https://doi.org/10 .1001/jama.1936.02770370020006.

13. Susan Reverby, *Examining Tuskegee: The Infamous Syphilis Study and Its Legacy* (Chapel Hill: University of North Carolina Press, 2009).

14. See Albert Jonsen, Oral History, interview by Bernard Schwetz, May 14, 2004, https://www.hhs.gov/ohrp/education-and-outreach/luminaries -lecture-series/belmont-report-25th-anniversary-interview-ajonsen/index .html.

15. National Commission for the Protection of Human Subjects of and Biomedical and Behavioral Research, "The Belmont Report: Ethical Principles & Guidelines for Research Involving Human Subjects" (Department of Health, Education, and Welfare, April 18, 1979), https://www.hhs.gov/ohrp /sites/default/files/the-belmont-report-508c_FINAL.pdf.

16. The report itself is accompanied by more than a thousand pages of appendices, explaining in detail their thinking on how to operationalize ethics and social norms in a way that could become a government-approved procedural specification.

17. Tom L. Beauchamp, *Standing on Principles: Collected Essays* (New York: Oxford University Press, 2010), 6.

18. Karen Lebacqz, interview by LeRoy Walters, October 26, 2004, https:// www.hhs.gov/ohrp/education-and-outreach/luminaries-lecture-series/ belmont-report-25th-anniversary-interview-klebacqz/index.html.

19. United States, ed., *Report and Recommendations: Institutional Review Boards*, DHEW Publication; No. (OS) 78-0008, 78-0009 (Washington, DC: US Department of Health, Education, and Welfare: for sale by the Supt. of Docs., US Govt. Print. Off, 1978).

20. Mike Monteiro, https://muledesign.com/2017/07/a-designers-code-of-ethics.

21. Jacob Metcalf, Emanuel Moss, and danah boyd, "Owning Ethics: Corporate Logics, Silicon Valley, and the Institutionalization of Ethics," *Social Research* 86, no. 2 (Summer 2019): 449–76.

22. Brent Mittelstadt, "Principles Alone Cannot Guarantee Ethical AI," *Nature Machine Intelligence* 1, no. 11 (November 2019): 501–7, https://doi.org/10 .1038/s42256-019-0114-4.

23. Inioluwa Deborah Raji et al., "Closing the AI Accountability Gap: Defining an End-to-End Framework for Internal Algorithmic Auditing," in *Proceedings of the 2020 Conference on Fairness, Accountability, and Transparency*, FAT* '20 (New York: Association for Computing Machinery, 2020), 33–44, https://doi.org/10.1145/3351095.3372873.

24. Shannon Vallor, "An Ethical Toolkit for Engineering/Design Practice,"

June 22, 2018, https://www.scu.edu/ethics-in-technology-practice/ethical
-toolkit/.

25. Metcalf, Moss, and boyd, "Owning Ethics."

26. Metcalf, Moss, and boyd, 465.

27. Theodore Vincent Purcell and James Weber, *Institutionalizing Corporate Ethics: A Case History* (New York: Presidents Association, Chief Executive Officers' Division of American Management Associations, 1979), 6; quoted in Ronald R. Sims, "The Institutionalization of Organizational Ethics," *Journal of Business Ethics* 10, no. 7 (July 1, 1991): 493, https://doi.org/10.1007/BF00383348.

28. Eric Johnson, "How Will AI Change Your Life? AI Now Institute Founders Kate Crawford and Meredith Whittaker Explain," Vox, April 8, 2019, https://www.vox.com/podcasts/2019/4/8/18299736/artificial-intelligence-ai-meredith-whittaker-kate-crawford-kara-swisher-decode-podcast-interview.

29. Ben Wagner, "Ethics as an Escape from Regulation: From 'Ethics-Washing' to Ethics-Shopping?," in *Cogitas Ergo Sum: 10 Years of Profiling the European Citizen*, ed. Emre Bayamlioglu et al. (Amsterdam University Press, 2018), 84–89, https://doi.org/10.2307/j.ctvhrd092.18.

30. Metcalf, Moss, and boyd, "Owning Ethics."

31. Henry T. Greely, "The Uneasy Ethical and Legal Underpinnings of Large-Scale Genomic Biobanks," *Annual Review of Genomics and Human Genetics* 8, no. 1 (2007): 343–64, https://doi.org/10.1146/annurev.genom.7.080505.115721.

32. Arvind Narayanan and Vitaly Shmatikov, "How to Break Anonymity of the Netflix Prize Dataset" (arXiv, November 22, 2007), https://doi.org/10.48550/arXiv.cs/0610105.

33. Pierangela Samarati and Latanya Sweeney, "Protecting Privacy When Disclosing Information: K-Anonymity and Its Enforcement through Generalization and Suppression," 1998.

34. Cynthia Dwork and Moni Naor, "On the Difficulties of Disclosure Prevention in Statistical Databases or The Case for Differential Privacy," *Journal of Privacy and Confidentiality* 2, no. 1 (September 1, 2010): 94, https://doi.org/10.29012/jpc.v2i1.585.

35. Cynthia Dwork, "Differential Privacy," in *Automata, Languages and Programming*, ed. Michele Bugliesi et al., Lecture Notes in Computer Science (Berlin, Heidelberg: Springer, 2006), 4, https://doi.org/10.1007/11787006_1.

36. Cathy O'Neil, *Weapons of Math Destruction: How Big Data Increases Inequality and Threatens Democracy* (New York: Crown, 2016); Virginia Eubanks, *Automating Inequality: How High-Tech Tools Profile, Police, and Punish the Poor* (New York: St. Martin's Press, 2017); Ruha Benjamin, *Race after Technology: Abolitionist Tools for the New Jim Code* (Cambridge, UK; Medford, MA: Polity Press, 2019).

37. Julia Angwin, Jeff Larson, Surya Mattu, Lauren Kirchner, "Machine Bias," ProPublica, May 23, 2016, https://www.propublica.org/article/machine-bias-risk-assessments-in-criminal-sentencing.

38. Arvind Narayanan, *Tutorial: 21 Fairness Definitions and Their Politics*, 2018, https://www.youtube.com/watch?v=jIXIuYdnyyk. See the notes at Shubham Jain, "TL;DS - 21 Fairness Definition and Their Politics by

Arvind Narayanan," July 19, 2019, https://shubhamjain0594.github.io/post
/tlds-arvind-fairness-definitions/.

39. Julie Zhuo, "How Do You Set Metrics?," *The Year of the Looking Glass*
(blog), August 10, 2017, https://medium.com/the-year-of-the-looking-glass
/how-do-you-set-metrics-59f78fea7e44.

40. Michael Kearns and Aaron Roth, *The Ethical Algorithm: The Science of Socially
Aware Algorithm Design* (New York: Oxford University Press, 2020), 78.

41. Will Douglas Heaven, "Predictive Policing Algorithms Are Racist. They
Need to Be Dismantled," *MIT Technology Review,* July 17, 2020, https://
www.technologyreview.com/2020/07/17/1005396/predictive-policing
-algorithms-racist-dismantled-machine-learning-bias-criminal-justice/.

42. Kearns and Roth, *The Ethical Algorithm*, 63.

43. Catherine D'Ignazio and Lauren F. Klein, *Data Feminism* (Cambridge, MA:
MIT Press, 2020), 61.

44. Matthew Le Bui and Safiya Umoja Noble, "We're Missing a Moral Framework of Justice in Artificial Intelligence," in *The Oxford Handbook of Ethics
of AI* (New York: Oxford University Press, 2020), 178, https://doi.org/10
.1093/oxfordhb/9780190067397.013.9.

45. Julia Powles and Helen Nissenbaum, "The Seductive Diversion of 'Solving' Bias in Artificial Intelligence," Medium, December 7, 2018, https://
medium.com/s/story/the-seductive-diversion-of-solving-bias-in-artificial
-intelligence-890df5e5ef53.

46. Frank Pasquale, "The Second Wave of Algorithmic Accountability,"
Law and Political Economy (blog), November 25, 2019, https://lpeblog
.org/2019/11/25/the-second-wave-of-algorithmic-accountability/.

47. Rodrigo Ochigame, "The Invention of 'Ethical AI': How Big Tech Manipulates Academia to Avoid Regulation," *The Intercept* (blog), December
20, 2019, https://theintercept.com/2019/12/20/mit-ethical-ai-artificial
-intelligence/.

48. Thao Phan et al., "Economies of Virtue: The Circulation of 'Ethics' in
Big Tech," *Science as Culture*, November 4, 2021, 7, https://doi.org/10
.1080/09505431.2021.1990875.

49. Shoshana Zuboff, *The Age of Surveillance Capitalism: The Fight for a Human
Future at the New Frontier of Power* (New York: PublicAffairs, 2019).

CHAPTER 12: PERSUASION, ADS, AND VENTURE CAPITAL

1. Herbert Simon, "Designing Organizations for an Information-Rich
World," in *Computers, Communications, and the Public Interest*, ed. Martin
Greenberger (Baltimore: Johns Hopkins Press, 1971), 40.

2. Paul Lewis, "Fiction Is Outperforming Reality: How Youtube's Algorithm
Distorts Truth," *The Guardian*, February 2, 2018, sec. Technology, http://
www.theguardian.com/technology/2018/feb/02/how-youtubes-algorithm
-distorts-truth.

3. "Easter Sun Finds the Past in Shadow at Modern Parade," *New York
Times*, April 1, 1929, https://timesmachine.nytimes.com/timesmach
ine/1929/04/01/95899706.pdf.

4. Edward L. Bernays, *Propaganda* (New York: H. Liveright, 1928).

5. Simon, "Designing Organizations for an Information-Rich World," 41.

6. Richard Serra, "TV Delivers People (1973)," *Communications* 48, no. 1
(1988): 42–44.

7. Neil Postman, *Amusing Ourselves to Death: Public Discourse in the Age of Show Business* (New York: Penguin Books, 1986).

8. For a deeper engagement with Postman and his forms of media determinism, see Siva Vaidhyanathan, *Antisocial Media: How Facebook Disconnects Us and Undermines Democracy* (New York: Oxford University Press, 2018), pp. 21–26.

9. https://www.w3.org/People/Berners-Lee/1991/08/art-6484.txt.

10. Ethan Zuckerman, "The Internet's Original Sin," *The Atlantic*, August 14, 2014, https://www.theatlantic.com/technology/archive/2014/08/advertising-is-the-internets-original-sin/376041/.

11. Michael H. Goldhaber, "The Attention Economy and the Net," *First Monday* 2, no. 4 (April 7, 1997), http://firstmonday.org/ojs/index.php/fm/article/view/519.

12. Goldhaber, "The Attention Economy and the Net."

13. Daniel Thomas and Shannon Bond, "BuzzFeed Boss Finds Natural Fit for Social Content on Mobile," *Financial Times*, March 14, 2016, https://www.ft.com/content/4f661ea8-e782-11e5-a09b-1f8b0d268c39.

14. Not all the prophecies of Goldhaber have yet to come to pass: he also prophesied a purely attention-based economy in which money would play no role. Although one can now barter mobile-phone minutes, or trade time in language lessons with another student in complementary languages, we do not yet have economies in which attention has fully replaced money.

15. Goldhaber, "The Attention Economy and the Net."

16. Stewart Brand, "The World Information Economy," *The Whole Earth Catalog*, no. Winter (1986): 88.

17. For a concise overview, see Christina Spurgeon, "Online Advertising," in *The Routledge Companion to Global Internet Histories* (Routledge, 2017); Joseph Turow, *The Daily You: How the New Advertising Industry Is Defining Your Identity and Your Worth* (New Haven, CT: Yale University Press, 2011).

18. See Tim O'Reilly, "What Is Web 2.0," September 30, 2005, https://www.oreilly.com/pub/a//web2/archive/what-is-web-20.html. For skepticism about the novelty of this shift, see Matthew Allen, "What Was Web 2.0? Versions as the Dominant Mode of Internet History," *New Media & Society* 15, no. 2 (March 1, 2013): 260–75, https://doi.org/10.1177/1461444812451567.

19. Nick Couldry and Joseph Turow, "Big Data, Big Questions | Advertising, Big Data and the Clearance of the Public Realm: Marketers' New Approaches to the Content Subsidy," *International Journal of Communication* 8 (June 16, 2014): 1714.

20. Kim Cleland, "Media Buying & Planning: Marketers Want Solid Data on Value of Internet Ad Buys: Demand Swells for Information That Compares Media Options," *Advertising Age*, August 3, 1998, S18; discussed in Turow, *The Daily You*, 61.

21. Cleland, "Marketers Want Solid Data"; discussed in Turow, *The Daily You*, 61.

22. Rick Bruner, " 'Cookie' Proposal Could Hinder Online Advertising: Privacy Backers Push for More Data Controls," *Advertising Age*, March 16, 1997, 16; discussed in Turow, *The Daily You*, 58.

23. Quoted in Meg Leta Jones, "Cookies: A Legacy of Controversy," *Internet Histories* 4, no. 1 (January 2, 2020): 94, https://doi.org/10.1080/24701475

.2020.1725852. See also David M. Kristol, "HTTP Cookies: Standards, Privacy, and Politics," *ACM Transactions on Internet Technology* 1, no. 2 (November 1, 2001): 151–98, https://doi.org/10.1145/502152.502153.

24. CNET News staff, "Ads Find Strength in Numbers," CNET, November 4, 1996, https://www.cnet.com/tech/tech-industry/ads-find-strength-in-numbers/.

25. See Jones, "Cookies," 95; Matthew Crain, *Profit Over Privacy: How Surveillance Advertising Conquered the Internet* (Minneapolis: University of Minnesota Press, 2021), 125–29.

26. Crain, *Profit Over Privacy*, 129.

27. Susan Wojcicki, "Making Ads More Interesting," *Official Google Blog* (blog), March 11, 2009, https://googleblog.blogspot.com/2009/03/making-ads-more-interesting.html.

28. Crain, *Profit Over Privacy*, 95.

29. Adam D'Angelo, *Quora*, 2010, https://www.quora.com/What-was-Adam-DAngelos-biggest-contribution-to-Facebook/answer/Adam-DAngelo.

30. Ashlee Vance, "In Ads We Trust," *Bloomberg Businessweek*, no. 4521 (May 8, 2017): 6–7.

31. John White, *Bandit Algorithms for Website Optimization* (O'Reilly Media, Inc., 2012); William R Thompson, "On the Likelihood That One Unknown Probability Exceeds Another in View of the Evidence of Two Samples," *Biometrika* 25, no. 3/4 (1933): 285–94.

32. For a nuanced account of Facebook's effects that rejects a simple demonization narrative, see Vaidhyanathan, *Antisocial Media*, e.g., at pp. 16–17.

33. James Grimmelmann, "The Platform Is the Message," SSRN Scholarly Paper (Rochester, NY: Social Science Research Network, March 1, 2018), https://papers.ssrn.com/abstract=3132758.

34. Zeynep Tufekci, "Engineering the Public: Big Data, Surveillance and Computational Politics," *First Monday*, July 2, 2014, https://doi.org/10.5210/fm.v19i7.4901.

35. Edward L. Bernays, "The Engineering of Consent," *The Annals of the American Academy of Political and Social Science* 250 (1947): 115.

36. Salman Haqqi, "Obama's Secret Weapon in Re-Election: Pakistani Scientist Rayid Ghani," DAWN.COM, January 21, 2013, https://www.dawn.com/2013/01/21/obamas-secret-weapon-in-re-election-pakistani-scientist-rayid-ghani/.

37. Rayid Ghani et al., "Data Mining for Individual Consumer Models and Personalized Retail Promotions," *Data Mining Methods and Applications*, 2007, 215.

38. Ethan Roeder, "I Am Not Big Brother," *New York Times*, December 6, 2012, http://www.nytimes.com/2012/12/06/opinion/i-am-not-big-brother.html?_r=0.

39. Zeynep Tufekci, "Yes, Big Platforms Could Change Their Business Models," *Wired*, December 17, 2018, https://www.wired.com/story/big-platforms-could-change-business-models/.

40. Tufekci, "Engineering the Public."

41. M. J. Salganik, *Bit by Bit: Social Research in the Digital Age* (Princeton, NJ: Princeton University Press, 2017), 10.

42. Mike Butcher, "Cambridge Analytica CEO Talks to TechCrunch about Trump, Hillary and the Future," *TechCrunch*, November 6, 2017, https://

social.techcrunch.com/2017/11/06/cambridge-analytica-ceo-talks-to
-techcrunch-about-trump-hilary-and-the-future/.

43. Trenholme J. Griffin, *A Dozen Lessons for Entrepreneurs* (New York: Columbia Business School Publishing, Columbia University Press, 2017), 146.

44. AnnaLee Saxenian, *Regional Advantage: Culture and Competition in Silicon Valley and Route 128, With a New Preface by the Author* (Cambridge, MA: Harvard University Press, 1996); Christophe Lécuyer, *Making Silicon Valley: Innovation and the Growth of High Tech, 1930–1970* (Cambridge, MA: MIT Press, 2006).

45. Josh Lerner, "The Government as Venture Capitalist: The Long-Run Impact of the SBIR Program," *The Journal of Private Equity* 3, no. 2 (2000): 55–78. Thanks to Ella Coon for stressing this.

46. Jerry Neumann, "Heat Death: Venture Capital in the 1980s," *Reaction Wheel* (blog), January 8, 2015, https://reactionwheel.net/2015/01/80s-vc .html.

47. Tom Nicholas, *VC: An American History* (Cambridge, MA: Harvard University Press, 2019).

48. Katrina Brooker, "WeFail: How the Doomed Masa Son-Adam Neumann Relationship Set WeWork on the Road to Disaster," *Fast Company*, November 15, 2019, https://www.fastcompany.com/90426446/wefail-how-the-doomed -masa-son-adam-neumann-relationship-set-wework-on-the-road-to-disaster.

49. Kai-Fu Lee, *AI Superpowers: China, Silicon Valley, and the New World Order* (Boston: Houghton Mifflin Harcourt, 2019).

50. Ryan Mac, Charlie Warzel, and Alex Kantrowitz, "Growth at Any Cost: Top Facebook Executive Defended Data Collection in 2016 Memo—And Warned that Facebook Could Get People Killed," *BuzzFeed News*, March 29, 2018, https://www.buzzfeednews.com/article/ryanmac/growth-at-any-cost -top-facebook-executive-defended-data.

51. Paul Lewis, " 'Fiction Is Outperforming Reality': How Youtube's Algorithm Distorts Truth," *The Guardian*, February 2, 2018, sec. Technology, http://www.theguardian.com/technology/2018/feb/02/how-youtubes -algorithm-distorts-truth.

52. Jane Jacobs, *The Nature of Economies* (New York: Modern Library, 2000).

CHAPTER 13: SOLUTIONS BEYOND SOLUTIONISM

1. Karl Manheim and Lyric Kaplan, "Artificial Intelligence: Risks to Privacy and Democracy," *Yale Journal of Law & Technology* 21, no. 1 (2019): 180, 181.

2. William H. Janeway, *Doing Capitalism in the Innovation Economy: Markets, Speculation and the State.* (Cambridge, UK: Cambridge University Press, 2012).

3. Marshall Kirkpatrick, "Facebook's Zuckerberg Says The Age of Privacy Is Over," *New York Times*, January 10, 2010, https://archive.nytimes.com/ www.nytimes.com/external/readwriteweb/2010/01/10/10readwriteweb -facebooks-zuckerberg-says-the-age-of-privac-82963.html.

4. Tim Cook, "We Believe That Privacy Is a Fundamental Human Right. No Matter What Country You Live in, That Right Should Be Protected in Keeping with Four Essential Principles," Tweet, *@tim_cook* (blog), October 24, 2018, https://twitter.com/tim_cook/status/1055035539915718656.

5. Blake Lemoine, "The History of Ethical AI at Google," Medium, May 17, 2021, https://cajundiscordian.medium.com/the-history-of-ethical-ai-at -google-d2f997985233.

6. Urooba Jamal, "An Engineer Who Was Fired by Google Says Its AI Chatbot Is 'Pretty Racist' and That AI Ethics at Google Are a 'Fig Leaf,' " *Business Insider*, July 31, 2022, https://www.businessinsider.com/google-engineer -blake-lemoine-ai-ethics-lamda-racist-2022-7.

7. danah boyd, "Where Do We Find Ethics?," *Medium*, June 15, 2016, https:// points.datasociety.net/where-do-we-find-ethics-d0b9e8a7f4e6; citing Audre Lorde, "The Master's Tools Will Never Dismantle the Master's House," in *Sister Outsider: Essays and Speeches* (Trumansburg, NY: Cross-ing Press, 1984), 110–14.

8. Anna Kramer, "How Twitter Hired Tech's Biggest Critics to Build Ethi-cal AI," Protocol—The people, power and politics of tech, June 23, 2021, https://www.protocol.com/workplace/twitter-ethical-ai-meta.

9. Michael Kearns and Aaron Roth, *The Ethical Algorithm: The Science of Socially Aware Algorithm Design* (New York: Oxford University Press, 2020), 16.

10. Kearns and Roth, 16.

11. Cynthia Rudin, "Stop Explaining Black Box Machine Learning Models for High Stakes Decisions and Use Interpretable Models Instead," *Nature Machine Intelligence* 1, no. 5 (May 2019): 10, https://doi.org/10.1038/s42256 -019-0048-x.

12. Annette Zimmermann, Elena Di Rosa, and Hochan Kim, "Technology Can't Fix Algorithmic Injustice," *Boston Review*, December 12, 2019, https://bostonreview.net/science-nature-politics/annette-zimmermann -elena-di-rosa-hochan-kim-technology-cant-fix-algorithmic.

13. Zimmermann, Di Rosa, and Kim.

14. Gina Neff et al., "Critique and Contribute: A Practice-Based Framework for Improving Critical Data Studies and Data Science," *Big Data* 5, no. 2 (June 2017): 85–97, https://doi.org/10.1089/big.2016.0050.

15. Mike Isaac, *Super Pumped: The Battle for Uber* (New York: W. W. Norton & Company, 2020), ch. 16.

16. Kate O'Flaherty, "Apple's Privacy Features Will Cost Facebook $12 Billion," *Forbes*, April 23, 2022, https://www.forbes.com/sites/ kateoflahertyuk/2022/04/23/apple-just-issued-stunning-12-billion-blow-to -facebook/.

17. Yochai Benkler et al., "Social Mobilization and the Networked Public Sphere: Mapping the SOPA-PIPA Debate," *Political Communication* 32, no. 4 (October 2, 2015): 594–624, https://doi.org/10.1080/10584609.2014 .986349.

18. For a powerful reflection on the continuing need for state regulation, see Frank Pasquale, *The Black Box Society: The Secret Algorithms That Control Money and Information* (Cambridge, MA: Harvard University Press, 2015).

19. Amy Kapczynski, "The Law of Informational Capitalism," *The Yale Law Journal* 129, n. 5 (2020), 1465.

20. Karl Manheim and Lyric Kaplan, "Artificial Intelligence: Risks to Privacy and Democracy," *Yale Journal of Law & Technology* 21 (2019): 162.

21. Michael Kearns and Aaron Roth, "Ethical Algorithm Design Should Guide Technology Regulation," *Brookings* (blog), January 13, 2020, https://www .brookings.edu/research/ethical-algorithm-design-should-guide-technology -regulation/.

22. Morgan Meaker, "Meta's Failed Giphy Deal Could End Big Tech's Spending

Spree," *Wired*, December 3, 2021, https://www.wired.com/story/facebook
-giphy-cma-global-template/.
23. Manheim and Kaplan, "Artificial Intelligence: Risks to Privacy and
Democracy," 186.
24. Quoted in Lina M. Khan, "Amazon's Antitrust Paradox," *The Yale Law
Journal*, 2017, 740.
25. Patrice Bougette, Marc Deschamps, and Frédéric Marty, "When Economics Met Antitrust: The Second Chicago School and the Economization of
Antitrust Law," *Enterprise & Society* 16, no. 2 (June 2015): 313–53, https://
doi.org/10.1017/eso.2014.18.
26. "General Data Protection Regulation (GDPR)—Official Legal Text," General Data Protection Regulation (GDPR), https://gdpr-info.eu/, art 22.
27. Meg Leta Jones, *Ctrl + Z: The Right to Be Forgotten* (New York: New York
University Press, 2016).
28. Khan, "Amazon's Antitrust Paradox."
29. Sarah T. Roberts, *Behind the Screen: Content Moderation in the Shadows of
Social Media* (New Haven, CT: Yale University Press, 2019).
30. Paul M. Barrett, "Who Moderates the Social Media Giants? A Call
to End Outsourcing" (NYU Stern—Center for Business and Human
Rights, June 2020), 4, https://static1.squarespace.com/static/5b6df958
f8370af3217d4178/t/5ed9854bf618c710cb55be98/1591313740497/NYU
+Content+Moderation+Report_June+8+2020.pdf.
31. Jeff Kosseff, *The Twenty-Six Words That Created the Internet* (Ithaca, NY:
Cornell University Press, 2019).
32. Jennifer S. Fan, "Employees as Regulators: The New Private Ordering in
High Technology Companies," *Utah Law Review*, Vol. 2019, no. 5 (2020): 55.
33. Alexis C. Madrigal, "Silicon Valley Sieve: A Timeline of Tech-Industry
Leaks," *The Atlantic*, October 10, 2018, https://www.theatlantic.com/
technology/archive/2018/10/timeline-tech-industry-leaks/572593/.
34. Daisuke Wakabayashi, "At Google, Employee-Led Effort Finds Men Are
Paid More Than Women," *New York Times*, September 8, 2017, https://
www.nytimes.com/2017/09/08/technology/google-salaries-gender-disparity
.html.
35. Catherine D'Ignazio and Lauren F. Klein, *Data Feminism* (Cambridge, MA:
MIT Press, 2020), 65.
36. Sarah Hamid, "Community Defense: Sarah T. Hamid on Abolish-
ing Carceral Technologies," *Logic Magazine*, August 31, 2020, https://
logicmag.io/care/community-defense-sarah-t-hamid-on-abolishing-carceral
-technologies/.
37. Ruha Benjamin, *Race after Technology: Abolitionist Tools for the New Jim
Code* (Cambridge, UK; Medford, MA: Polity Press, 2019).
38. Zimmermann, Di Rosa, and Kim, "Technology Can't Fix Algorithmic
Injustice."
39. Zimmermann, Di Rosa, and Kim.
40. Amy Kapczynski, "The Law of Informational Capitalism," 1460.
41. On norms, laws, architecture and markets, see, e.g., Lawrence Lessig, "The
New Chicago School," *The Journal of Legal Studies* 27, no. S2 (1998): 661–91.

INDEX

Abbate, Janet, 104, 146
Accenture, 288
Agar, Jon, 132–33
Aged Poor in England and Wales, The (Booth), 59
AI. *See* artificial intelligence
algorithms. *See* computational techniques; data; data analysis; machine learning; neural networks; pattern recognition; regression
Allen, Frances, 113–14
Alston, Philip G., 233
Amazon, 4, 215–16
Amazon's Mechanical Turk, 189, 220
American Economic Association, 54
American Medical Association, 95
Amusing Ourselves to Death (Postman), 259–60
Anthropometric Laboratory (Galton), 41, 42
antitrust regulation, 296–99
Aristotle, 51
artificial intelligence (AI), xii
  contemporary meaning of, 120
  Dartmouth Workshop and, 127–31, 134, 173
  data-driven approach to, 138–40
  digital computers and, 127–28, 129, 139–40
  ethics and product development in, 233–34, 245–46
  expert systems, 135–36, 137–38, 182–83
  foundations in field of, 127–28
  human knowledge and, 131–33
  human labor and, 219–21, 274
  knowledge acquisition bottleneck, 136–38, 274
  McCarthy and, 126–28, 133, 134, 135, 182
  multi-layer neural network and, 177–79, 185–86, 188
  naming of, 126–27
  real-world application metrics optimization and, 172–74, 194–95
  as redefined machine learning, 190–91
  research funding of, 133–35, 182
  rules vs. data in, 124–26, 134, 183
  symbolic approaches and, 123–24, 126–32, 134, 137–38, 175–79
  symbolic opposition and, 180, 183, 184, 185, 221
  Turing and, 121–23
  See also computational statistics
AT&T's Bell Labs. *See* Bell Labs
Atomic Energy Commission, 115n
attention economy
  algorithmically optimized UGC and, 264–65
  broadcast television business model and, 258–60
  internet advertising, origins of, 265–68
  optimizing engagement, 269–70
  venture capital and, 280–83
  World Wide Web and, 260–64
Automating Inequality (Eubanks), 9

backfire effect, 274
Baker, Erica, 302

Baran, Paul, 162
Barocas, Solon, 251–52
Bayes, Thomas, 109–10
Bayesian methods
  Bayes' rule, 108–9, 110–11
  industrial data analysis and, 103,
    111–12
  statistics, 107–12, 116–17, 181
bell curve, ix, 26, 31, 32, 38–39, 73
Bell Labs, xii
  accumulation of data in real-world
    systems and, 206
  after World War II, 120
  communications data and, 118
  data science and, 207, 208, 224
  ensemble modeling and algorithmic
    prediction, 188, 193–94
  exploratory data analysis and, 203–5,
    207
  future luminaries of applied compu-
    tational statistics and, 112
  kernel machines and, 187
  machine learning international
    researchers and, 184–85, 187
  machine learning and neural net-
    works, 171–72, 183, 185
  pattern recognition and, 179–80
  See also NSA
Belmont Report, The, 233, 236, 238–43,
    344n16
Bengio, Yoshua, 185
Benjamin, Ruha, 7, 251, 304
Bentham, Jeremy, 241
Benzécri, Jean-Paul, 183–84
Bernays, Edward, 257–58, 261,
    275–76
Berners-Lee, Tim, 260
big data
  age of, 5
  distributed database platform and,
    216–17
  electronic data processing transfor-
    mations, 146, 148–49
  high-dimensionality of, 209
  infrastructure and industrial-
    intelligence partnerships, 144–46,
    148–49, 217
  mass private industry data collection
    and, 142–43, 155–56, 166–67,
    224–25

technological integration and,
    149–50
technology-driven systems commer-
    cialization, 141–42
  See also NSA; privacy and justice
BigTable, 214, 217
biology, 43, 48, 57, 59, 199, 206
Biometric Laboratory, 44
biometric sciences, 50–53, 57
Biometrika (Pearson), 83
Black Americans, 54, 56–57
"Black Friday," 113
Bletchley Park, 102–4, 106, 106, 107,
    111, 121, 139
blitzscaling, 281
body mass index, 19
bombes, 103, 105
Booth, Charles, 59, 62
Bork, Robert, 299
Bosworth, Andrew, 283
Botstein, David, 210
Bottou, Léon, 187
Bouk, Dan, 14
boyd, danah, 5, 245, 287
Brand, Stewart, 263
Brandeis, Louis, 299
Breiman, Leo, 98, 187, 222–23
Brin, Sergey, 211, 212, 213, 214, 228,
    268
Brown, Patrick, 210
Browne, Simone, 14
Brunsviga mechanical calculator, 64
Bryan, Jennifer, 226
Buchanan, Bruce, 136
Buolamwini, Joy, 233
Burge, W. Lee, 160
Burke, Colin, 113

Cambridge Analytica, 9, 277
Campaigne, Howard, 103
capitalism, 145
Carpenter v. United States, 169
Carson, John, 73
Casilli, Antonio, 219–20
caste difference, 52
Caughey, Catherine, 104
causation, 49, 59–60, 64–65
Cerf, Vint, 5
Chambers, John, 205–6, 223
Chaslot, Guillaume, 257, 283

China Intelligence agencies, 4
Chowdhury, Rumman, 288
Chun, Wendy, 5
classism, 37, 50, 53
Cleveland, William, 196, 207–8, 223
cloud-hosted computing, 224–25
cognitive science, 130
Cohen-Cole, Jamie, 130
Cohn, Bernard, 14
Cold War, 144–46, 196, 197, 218, 276
Colossus, 103–4, *106*, 139
Communications Decency Act (1996),
   299
COMPAS, 251
computational social science, 243
computational statistics
   applied, 107, 112, 113, 116, 126, 140
   data-based predictions focus of,
      179–80
   industrial-intelligence resourcing
      and, 185, 216–19
   NSA and, 112, 168–69
   in opposition to mathematical statis-
      tics, 183–84, 221
computational techniques
   cryptographic Bayesian analysis and,
      107–12, 116
   data mining, 107, 162, 172–74,
      209–11, 212–13, 214–15, 228–29
   electromechanical computing devices
      and, 103, 105
   historical machine calculation, xii,
      44, 52, 64, 75, 105
   printed tables and, 46
   tabulators, 105
"Computer Advances Pioneered by
   Cryptologic Organizations" (Snyder), 115*n*
computerized data processing, 103–4,
   113–14, 117–18, 144–46, 148–49
computerized data storage technologies
   attention economy and, 258–60,
      269–70
   challenges of, 117–18
   industry data collection and, 119
   intelligence/industrial/academic
      complex and, 217
   NSA development support of,
      113–16, 144–45, 202
   pattern recognition and, 139–40

security and labor costs of, 146
   Turing on, 121–23
computers, digital
   AI and, 127–28, 129, 139–40
   EDP applied transformations, 146,
      148–49
   logic and math vs. data and profit, 119
   mathematical statistics and, 223
   neural networks and, 177–80, 183,
      185–91
   NSA and, 143–45
   parallel computers and, 186
   scale in collecting, processing, and
      storing data, 145
computers, origins of, 103–4, 123–24
   Bletchley Park researchers and, 102–4
   compilers, 133
   corporate data collection and, 119,
      142–43, 145
   industry adoption of, 142, 144–45,
      148–50
   NSA technological development of,
      143–45
   programming languages, 133
   Turing and, 121–23
"Computing Machinery and Intelligence" (Turing), 121–23
consumer credit industry, 149, 315*n*28
Cook, Tim, 286, 292
Coolidge, Calvin, 275
corporate power, xii, 305–7
   attention economy/venture capital
      consequences and, 280–83
   data-empowered algorithmic systems
      realignment of social order and,
      6–10
   data-empowered technology distribu-
      tion of, 8, 145–46
   design choices and misdirection in,
      290–91
   ethical principles and, 243–48,
      286–90
   hardware to information platforms
      technology shifts and, 160*n*
   numerical accountability used for,
      16–17
   origins of statistics and, 22–25
   personal data collection and inter-
      pretation history and, 22, 143,
      150–60

corporate power (*continued*)
  relationship to data profitability and, 285–86
  self-regulatory organizations and, 293–94
  state power as regulator and, 294–96
  technological threats to individuals and, 13–14
  technology ecosystem contests and, 291–93
  See also people power; state power
correlation
  dangers of reasoning from, 64–65
  Galton and, 40–43
  Pearson and, 47, 49, 60
  reification and, 70
  Spearman and, 68–69, *69*
  Yule and, 63, 64
COVID-19, 108–9, 226, 262
Crain, Matthew, 164, 267, 268
Crawford, Kate, 5
cryptography
  Bayesian analysis and, 107–12, 116
  digital computation scaling and, 142–43, 144
  pattern recognition and, 139–40
  power dynamics and, 105–7
  storage and, 114

D'Angelo, Adam, 269
Danziger, Kurt, 41, 42
DARPA (Defense Advanced Research Projects Agency), 134, 138, 170
"Dartmouth Summer Research Project on Artificial Intelligence, The" (McCarthy), 126–28
Dartmouth Workshop, 127–31, 134, 173
Darwin, Charles, 33, 35, 36
data (data-driven algorithmic decision-making systems), x, xiii
  activism against, 11
  advances of, 12–13
  algorithmically optimized UGC, 264–65
  applied ethics of IRB and, 234–36, 240–46
  Belmont report and, 238–40
  content moderation and, 274
  dangers of, 7–10

defining the field of data science, 196–201
disinformation concerns and, 10
early networking technologies and, 142, 145, 156, 159
engagement oriented platforms, 270–73
ethics and politics of, xi, 8–9, 188, 195, 243–48
future potential of, 306–7
history and, 14–17, 152
human labor and, 13–14, 146, 219–21, 274
law enforcement and, 10
libertarian political views of, 6, 119, 153, 163, 164, 166
methods of auditing and analyzing, 56
political persuasion architecture and, 274–78
predictive ensemble models, 187–91
predictive systems and, 179–80
real-world data metrics optimization and, 169–74, 194–95
reproduction of systemic inequalities and, 7–8
responsibilities of, 12–14
secret, proprietary, opaque processing in, 16, 255–56
technological fixes in ethics, fairness, and privacy, 249–54
World Wide Web and, 212–15, 260–64
data analysis
  Bayesian methods and, 103, 107–12, 116–17, 181
  data mining and, 107, 162, 172–74, 209–11, 212–13, 214–15, 228–29
  factor analysis, 70
  greater statistics and, 205–7, 225–26
  historical mathematical statistics, 23–25, 59–60, 103
  implications in numerical differences, 52–53
  industrial-scale statistics, 105, 107
  mathematically focused, theoretical statistics, 203
  mathematical statistics regression and, 39–40
  NSA's aggregate data analytical tools and, 168–69

paper to computer, 98, 103, 118–19, 202–8
sequential analysis, 96
S system for, 207
statistical analysis of differences and, 27–28, 226–27
statistical and computational tools for, 204
systematic profile analysis and, 73–75, 149, 150–51
technological challenges in, 208
See also data science; Pearson, Karl; Quetelet, Adolphe
databases, 148–49, 152, 155–60, 213–14, 216–17
data collection
big data collection scale and, 142–43, 155–56, 166–67, 224–25
corporate data collection, 119, 141, 142–43, 145, 161–62
corporate data mining, 107, 162, 172–74
de facto national database and, 152, 155–59
digital computers historical scale of, 145
institutionalized data usage/retrieval and, 111–12, 113–18, 119, 148–49, 152–53, 160–61
limitations on, 169
normalization of, 145–46, 159–60
numerical data collection methods, 23, 41, 43–44, 73–75
systematic profile analysis and, 73–75, 149, 150–51
Data Feminism (D'Ignazio and Klein), 219, 227, 303
data journalism, 227
data mining, 107, 162, 172–74, 209–11, 212–13, 214–15, 228–29
Data Paper for Heredity Investigations (Pearson), 45
data processing, computerized origins of, 103–4, 113–14, 117–18, 144–46, 148–49
"Data Science: An Action Plan for Expanding the Technical Areas of the Field of Statistics" (Cleveland), 196

data science
advertising and, 270–73
applied ethics of IRB process and, 234–36, 240–46
Bell Labs and, 207, 208, 224
Cleveland and, 207–8
competitive common task framework and, 193–94
data-driven racism and, 50
data mining, 107, 162, 172–74, 208–11, 212–13, 214–15, 228–29
defining roles of, 196–201
development of new architectures and, 214
early use of key algorithms in, 180
ethics and, 228–29, 233–36, 238–40, 241–46, 247–48, 249–54
exploratory data analysis and, 202–8
history of, xi, 48–49, 52–53, 224–26
human labor and, 146, 219–21, 274
industrial, 197, 200–201
industrial-intelligence resourcing and, 185, 216–19
industrial-scale machine learning, 195
pseudoscientific physiognomy and, 226–27
real-world application metrics optimization and, 172–74, 194–95
real-world statistics and, 221–24
research and academic adoption of, 225–28
statistics vs real-world data, 199–201
subject humanity and, 6, 238–40
technology tools and, 227–28
World Wide Web and, 212–15
See also computational statistics
David, Florence Nightingale, 46, 83
deep learning, 176, 186, 188, 189, 190
Deming, W. Edwards, 96
DENDRAL, 136
Desrosières, Alain, 39, 59, 67
Dick, Stephanie, 133
Didier, Emmanuel, 14, 74–75
differential privacy, 250
digital computation, xii, 44, 52, 64, 75, 105
D'Ignazio, Catherine, 219, 227, 253, 303
DiResta, Renée, 9

Di Rosa, Elena, 290–91, 304
Divine Providence, 25
Doctorow, Cory, 278
Donnelly, Kevin, 25
Donoho, David, 194
Doriot, George, 279
DoubleClick, 266, 267
Dryer, Theodora, 90
Du Bois, W. E. B., 50–51, 53, 55–56
Du Boisson, Dorothy, *106*
Duda, Richard, 140
Dumbill, Edd, 141
"During World War II" (Neyman), 97
Dwork, Cynthia, 250
Dyson, Esther, 165

Eachus, Joseph, 115*n*
Edwards, Paul, 149
Efron, Bradley, 223
Elderton, Ethel, 44
electromechanical computing devices
    (bombes), 103, 105
electronic data processing (EDP), 148
empirical techniques, ix, 24, 52, 93,
    125, 194
Engineering Research Associates
    (ERA), 144
Enigma machine, 103, 105
Enlightenment, 21, 22, 28
Ervin, Sam, 151
Ethical Algorithm, The (Kearns and
    Roth), 253, 289
ethnological evidence, 53
Eubanks, Virginia, 8, 9, 251
Euclid, 93
eugenics
    biometric sciences and, 50–51
    Fisher and, 87–88
    Galton on, 37, 70–71*n*
    historical significance of, 50
    historical view of, 38
    intelligence testing and, 68–72
    natural selection and, 36–37
    normal curve racial ranking and,
        38–39
    Pearson and, 46–48
Eugenics Lab, 44
expert systems, AI and, 135–36,
    137–38, 182–83
exploratory data analysis, 202–8

Facebook
    ad model, 267
    AI ethicists, 7, 11
    applied ethics of IRB process and,
        234–36, 240–46
    business model of, 4
    Cambridge Analytica and, 9
    challenges of scale and, 215–16
    emotional contagion study and,
        234–36
    Fairness Flow, 288
    Hadoop, 216
    moderators and, 274
    persuasion architecture feature
        "lookalike audiences," 274–75
    privacy and, 160
    venture capitalism and, 282
factor analysis, 70
fairness/accountability/transparency
    concerns
    algorithmic systems and, 16–17,
        251–52, 253, 254–56
    Belmont principles and, 241–43
    data science incentives and, 6, 234
    Facebook Fairness Flow and, 288
    movement for, 6–7, 246–48
    new computing technologies and
        labor costs, 146, 219–21, 274
    new technology design decisions
        and, 149–50
    optimizing objectives and, 252–53
    people power co-optation of, 7
    "Principles for Accountable Algo-
        rithms" on, 11–12
    quantifying fairness and, 251–53,
        289–91
    technical approaches to, 288–89
    technological fixes in ethics, fairness,
        and privacy, 249–54
    technology company processes and,
        11, 243–45
    Wallach on, 3–4, 250
Fan, Jennifer, 301
Fayyad, Usama, 208
Federal Trade Commission (FTC), 235
Feigenbaum, Edward, 136
Felten, Edward, 168
*First Census Optical Character Recogni-
    tion System Conference, The,* 171
First Monday (Goldhaber), 261

First Universal Races Congress (1911), 50–51, 53
Fisher, Ronald, 79, 83–89, 92, 93, 107
    Fisher, R. A., 269
Five Eyes, 106–7
Food and Drug Administration (FDA), 95
Foreign Intelligence Surveillance Court (FISC), 167
Foucault, Michel, 14, 284
Fourteenth Amendment, 55
Fourth Amendment, 168
Frederickson, George, 56
Freedman, David, 65, *66*
French Revolution, 19
Friedman, Milton, 163
"From Association to Causation: Some Remarks on the History of Statistics" (Freedman), *66*

Galton, Francis, *40, 42*
    data correlation and, 40–43
    on human quality improvement, 36–37
    on natural ability, 70–71
    on natural equality, 37–38
    ranking/classifying peoples and races and, 38–39, 73
    "science of individual differences" and, 33, 34, 71
    statistical regression and, 39–40
Gandy, Oscar, Jr., 5, 164
Gauss, Carl, 26
Gaussian curve, ix
GCHQ (Government Communications Headquarters) (United Kingdom), 6, 112, 166
Gebru, Timnit, 7, 233, 234, 246
Gekko, Gordon, 284
General Data Protection Regulation (GDPR), 297
genetic engineering, 243
German cyphers, 102–3
Ghani, Rayid, 276
Gibson, William, 8
Ginsberg, Allen, 196–97
Gitelman, Lisa, 21
global climate change, 58
Global War on Terror, 217
Goldhaber, Michael, 261, 262, 263

Goldwater, Barry, 151
Good, Jack, 112, 139
Google
    ads and, 263
    AI ethicists, 7, 11, 233–34, 247, 253
    analyzing real-world data, 4
    BigTable, 217
    challenges of scale and, 215–16
    ethics and, 246–47
    Facets, 288
    MapReduce, 216
    PageRank, 214, 263
    search engine creation, 214
    search feature and, 5–6, 17, 211, 212
    surveillance advertising and, 267
    training neural nets and, 190
    venture capitalism and, 282
    What-If Tool, 288
Gosset, William, 79–83, 86, 91–92
Gould, Stephen Jay, 72
Grammar of Science (Pearson), 90, 91
graphics processing units (GPU), 190
Gray, Mary L., 219–20
greater statistics, 205–7, 225–26
Grimmelmann, James, 273
"Growing Up with Computers at NSA (Top Secret Umbra)," 116*n*
Grus, Joel, 197–98
Guinness, Edward, 78
Gurley, Bill, 279
Guyon, Isabelle, 172, 185

Hacking, Ian, 19, 22, 28, 31, 32
Haitian Revolution, 19
Hamid, Sarah T., 303–4
Hammerbacher, Jeff, 196, 197, 216, 224, 269
Hardt, Moritz, 251–52
Hart, Peter, 140
Hayashi Chikio, 184
Hereditary Genius (Galton), 38, 70–71*n*
"Hereditary Talent and Character" (Galton), 36
Heyck, Hunter, 129
Hicks, Mar, 104
HIPAA (1996), 155
Hobbes, Thomas, 28
Hodes, Martha, 14
Hoffman, Frederick, 54–55, 56, 57, 59, 76

Hollerith punched card machines, 75, 169
homme moyen (average man), 27
Hopper, Grace, 133
Hotelling, Harold, 96, 97, 203
Hume, David, 109–10
Hutson, Jevan, 227
Hwang, Tim, 278
hypothesis testing significance levels
  cookbook approach, 93–96
  Fisher and, 79, 84–86
  Gossett and, 79, 81–83
  Neyman and, 79, 89, 91–93

IBM, 7, 11, 288
IBM card-processing machines (tabulators), 105
Igo, Sarah, 14
immigration, 71, 75
industrial-academic complex, 11, 200, 217
Industrial Revolution, 13
inequality
  science of, 54–56
  socioeconomic, 9–10, 50, 57
  structural, 7–8, 253–54, 302
information processing systems, 130
institutional review board (IRB), 234–36, 240–46
intelligence testing, 67–70, 71–72, 73
International Health Exhibition (1884), 41
internet, 159, 162, 164–66, 224, 264–68, 299–301
  See also World Wide Web
inventive technology/new science outcomes
  academic "capture" and, 10–11, 254–55
  adtech and persuasion architectures and, 277–78
  COMPAS algorithm of criminal recidivism and, 251
  data mining results and, 228–29
  emotional contagion study and, 234–36
  Google search and, 5–6
  large data collection/practical analysis and, 206–7

obfuscation of laborers worldwide, 146, 219–21, 274
"Project Match" and categories of people, 158–59
Quetelet's average man, 27
science of biology and, 199
socioeconomic, sexual, and racial disparities of algorithmic systems, 253–54
"Strategic Subjects List" and, 9–10
technological determinism and, 14, 306
venture capital and new products, 279–83
  See also big data; eugenics, biometric sciences; fairness/accountability/transparency concerns; NSA; racism; scientific racism; surveillance capitalism
"Investigation into the Causes of Changes in Pauperism in England, An" (Yule), 60–64
Irani, Lilly, 220, 221
IRB. *See* institutional review board
Ireland, Eleanor, 104
Isaac, Mike, 292
Israel Intelligence agencies, 4

Jackman, Molly, 235
Jacob, S. M., 44
Jacobs, Jane, 283
Janeway, William, 284–85
J. D. Rockefeller's Standard Oil Company, 296
Jim Crow, 55
Jordan, Michael, 191, 289

Kanerva, Lauri, 235
k-anonymity, 250
Kant, Immanuel, 241
Kapczynski, Amy, 294, 305
Kaplan, Lyric, 284, 295, 296
Kearns, Michael, 253, 289, 295
Kefauver-Harris Amendment (1962), 95
Kenyon, David, 103
kernel machines, 187
key performance indicator (KPI), 194
Khan, Lina, 298, 299
Kim, Hochan, 290–91, 304

Klein, Lauren, 219, 227, 253, 303
knowledge acquisition bottleneck, AI,
  136–38, 274
Knowledge Discovery in Databases
  (KDD), 209
Kranzberg's First Law, 3, 12
Kullback, Solomon, 105, 118
Kullback-Leibler divergence, 117
Kurzweil, Ray, 290

Langley, Pat, 175, 192
Laplace, Pierre-Simon, 26
Lauer, Josh, 149
law of errors, 25–27, 30, 32, 39
Lévi-Strauss, Claude, 125
Lebacqz, Karen, 242
Le Bui, Matthew, 254
LeCun, Yann, 172, 185, 187
Lederberg, Joshua, 136
Lee, Alice, 44, 47
Lee, Kai-Fu, 282
Lehmann, Erich L., 92
lesser statistics, 205
Li, Fei-Fei, 188
Li, Xiaochang, 185
Lighthill, James, 134, 182
LinkedIn, 215–16
Lorde, Audre, 287
low-rights environments, people in, 8,
  159

Machiavelli, Niccolò, 51
machine learning
  Ad-supported UGC-hosting sites,
    264–65
  advanced technologies and, 214
  analyzing real-world data and, 3–4
  Bell Labs and, 171–72, 183, 184–85,
    187
  common task framework and,
    193–94
  data mining and, 107, 162, 172–74,
    209–11, 212–13, 214–15, 228–29
  data science academics and, 199–201
  human labor and, 219–21
  industrial-intelligence ties and,
    139–40
  industrial-scale machine learning,
    195
  NSA and, 117, 216–17

numerical prediction and classifica-
    tion, 175, 191–92
  optimized advertising, 270–73
  pattern recognition fundamentals
    and, 139–40, 175, 181, 183
  pattern recognition origins, 179–80
  power of platforms, 169
  practical engineering traditions of,
    181, 183
  predictive ensemble models and
    neural nets, 187–91
  problem solving emulation and,
    137–38
  pseudoscientific physiognomy,
    226–27
  real-world application metrics opti-
    mization and, 172–74, 194–95
  redefinition of, 190–91
  sponsored industrial research and,
    10–11
  See also artificial intelligence
Madigan, David, 223
Mahalanobis, Prasanta C., 35, 51–53,
  67–68, 74
Mallock's Machine, 52
Manes, Stephen, 160n
Manheim, Karl, 284, 295, 296
Mansfield Amendment (1969), 134
MapReduce, 214
master codebreakers, 103
mathematical rigor
  Bayesian analysis and, 102, 103, 116
  benefits of, 15
  biometrical data and, 51–52
  data science and academic statistics,
    222–23
  engineering and, 113
  mathematical statistics field and, 97
  mathematical statistics profession
    and, 98
  Pearson and, 43
  Quetelet's data abstraction and, 27
  statistical effort and, 93–94
  Tukey and, 203–4
mathematical statistics, x, xii
  alternative statistical models, 183–84
  Breiman and, 222–23
  historical data-driven analysis,
    23–25, 59–60, 103
  infrastructure, 44, 46, 64, 75

mathematical statistics (*continued*)
  level of significance and hypothesis
    testing, 79, 81–83, 84–86, 89,
    91–96
  Mahalanobis and, 35, 51–53, 74
  Neyman and, 97–98
  numerical accountability methods
    and, 15–17
  numerical data collection methods,
    23, 41, 43–44, 73–75
  numerical data methodological his-
    tory, 21–24, 38–40
  quantification of peoples, 14–15,
    22–23
  Quetelet, Pearson's vision and, 43
  ranking/classifying individuals,
    41–43
  regression analysis, 39–40
  World War II and, 96–98, 124, 125
Mao Zedong, 75
McCarthy, John
  artificial intelligence and, 126–28,
    133, 134, 135, 182
  criticism of, 131, 132
  machine intelligence and, 124
McCarthy, Kathleen, *157*
McLuhan, Marshall, 273
McNealy, Scott, 160
Merchant, Emily, 14
Mertia, Sandeep, 74
META (Machine Learning, Ethics,
    Transparency and Accountabil-
    ity), 187, 288
metadata, 166, 167, 168
Metcalf, Jacob, 245, 248
Michie, Donald, 112, 137
Microsoft, 3, 288
microtargeting, 9, 265–68, 276–78
MIDAS (Mining Data At Stanford), 211
militarism, 145
military. *See* NSA; Office of Naval
    Research
Mill, John Stuart, 54, 76, 241
Miller, Arthur, 151
Miller, Kelly, 76
Minsky, Marvin, 132, 135
Mitchell, Margaret, 7, 233, 234, 246,
    287
Mitchell, Tom, 191, 289
Mittelstadt, Brent, 245

Moody, Juanita, 101
Moss, Emanuel, 245
Muhammad, Khalil Gibran, 14, 57
Muigai, Wangui, 73, 74
multi-layer neural network, AI and,
    177–79, 185–86, 188
Musk, Elon, 290
MYCIN, 136

Nakamura, Lisa, 3, 8
Naked Society, The (Packard), 150
Napoleonic Empire, 19
Narayanan, Arvind, 249, 251
National Institute of Standards and
    Technology (NIST), 170, 171, 173
National Science Foundation (NSF), 98
National Security Agency (NSA). *See*
    NSA (National Security Agency)
natural selection, 36–37
Nature of Economies, The (Jacobs), 283
Neff, Gina, 225–26, 291
neo-Brandeisian antitrust regulation,
    299
Netflix, 192–93, 194
Neuman, Jerry, 280
Neumann, Adam, 281
neural networks, 177–80, 183, 185–91
Newell, Allen, 129–31, 133
Newman, Max, 139–40
Neyman, Jerzy, 89–93, 97, 107
Nicholas, Tom, 280
Nightingale, Florence, 18, 33–34, 36,
    73
9/11, 6, 166
Nissenbaum, Helen, 5, 254, 290
Nix, Alexander, 277
Nobel Prize in Economics, 130
Noble, Safiya, 5–6, 254, 290
Noonan, Peggy, 20
normal curve, ix, 26, 31, 32, 38–39, 73
Nozick, Robert, 163
NSA (National Security Agency)
  after World War II, 113–18
  aggregate data analytical tools and,
    168–69
  before big data, 101
  Cold War state-driven capitalism
    and, 144–46
  communications surveillance capa-
    bility expansion and, 6, 166–67

computational statistics at, 112,
168–69
computer technological developments and, 113–16, 143–45, 202
data mining and, 214–15
digital computers and, 143–45
distributed database platform and,
216–17
early data processing and storage
needs of, 113–16
industrial scale Bayesian analysis
and, 103, 111–12
institutional culture of, 218–19
large data sets and, 202, 206, 208
machine learning and, 117, 216–17
metadata vs. content surveillance
and, 167–69
paper to computer data analysis, 98,
103, 118–19, 202–8
pattern recognition technologies and,
139–40, 175, 179–80
See also Bell Labs
null hypothesis, 81, 85, 89
numerical accountability, 15–17, 22,
31, 151–52, 155–56, 159–60
Nunn, Sam, 153
NVIDIA, 190
Nye, Robert, 50

Ochigame, Rodrigo, 255
O'Conner, Kevin, 266–67
Office of Naval Research, 96, 97, 98,
117, 134, 214–15
O'Neil, Cathy, 6, 199, 251, 288
"Operation Vengeance," 105–6
O'Reilly, Tim, 264
Origin of Species (Darwin), 36

Packard, Vance, 150
Page, Larry, 211, 212, 213, 214, 268
PageRank, 214
Papert, Seymour, 135
Pasquale, Frank, 11, 254
Patil, Dhanurjay "DJ," 197
PATRIOT Act (2001), 166
*Pattern Classification and Scene Analy-
sis* (Duda and Hart), 140
pattern recognition, 139–40, 175,
179–80, 181, 183, 224–25
Pearson, Egon, 89, 91, 92

Pearson, Karl, *45*
about, 43, 107
on correlation, 47, 49, 60
data collection methods, 43–44
data and machine calculations, 44,
46
eugenic socialism and, 46–48
on knowledge, 91
Seal on, 51
on social and biological sciences,
48–49
on Spearman's hypothesis of general
intelligence, 72
systematic data collection and analysis, 71, 73
Penn, Jonnie, 127, 130
people power, 305–6
algorithmic accountability and,
250–54
algorithmic decision systems dangers
to, 10, 152–53, 155, 158–60
collective action publicity, 301–2
data-empowered capabilities effect
upon, 13–14
numerical accountability used for,
15–16
shareholder activism as, 302
tech employees and, 302–3
user public engagement or disengage-
ment and, 304–5
See also corporate power; state
power
People's Republic of China, 75
Perceptron, 177, 179
Peretti, Jonah, 261
Phillips, Christopher, 94
physiognomy, 226–27
Pigou, Arthur, 64–65
Pólya, George, 130
Poon, Martha, 149, 315n28
Porter, Theodore, 15, 32, 43, 95
Postman, Neil, 259, 273
poverty
Booth and, 59
correlation and causation, 64–65
multiple regression and, 58
pauperism reification and, 66–67
state policies and, 58–59
Yule and, 59–64, 65
Powles, Julia, 254

predictive (discriminative) models of induction, 184
"Principles for Accountable Algorithms," 11–12
Privacy Act (1974), 155–56
Privacy Journal (Smith), *157*
privacy and justice
Belmont report and, 238–43
big data collection scale and, 142–43, 155–56, 166–67, 224–25
big data and emaciated social views of, 163–69
data justice and, 253–54
de facto national database collection, 152, 155–59
differential privacy, 250
discriminatory gatekeeping and, 153
internet commercialization and, 164–65
legal/technological analysis and, 6–7, 163–65, 166–69
legislation for, 154–56
personal computers and, 159–60
Privacy Protection Study Commission and, 143, 155
privacy vs. free circulation of information, 152–54, 161–62
revival of privacy, 150–53
technological fixes in ethics, fairness, and privacy, 249–54
value of information, 152–54, 160, 161
Zuckerberg on, 160
Privacy Protection Study Commission, 143, 155
"Private Lives? Not Ours!" (Manes), 160*n*
Propaganda (Bernays), 258
Purcell, Theodore, 247
p values, 84, 94

Quetelet, Adolphe, ix
about, 19–20, 73
"average man" data abstraction and, 19, 26–28
data journal publications and, 25
on data regarding crime, 29–30
on discerning the nature of humanity, 28–29
influence of, 32–34

as inspiration for Florence Nightingale, 18
social physics and, 18–20, 26–28, 29–32
statistical analysis and, 24, 25–26
Quinlan, J. Ross, 137

Race After Technology (Benjamin), 304
racial criminalization, 57
racial sciences, 54–57
See also scientific racism
racism, scientific. *See* scientific racism
racism
eugenics and, 50, 51
Google search and, 5–6, 17
invasions of privacy and, 153
Jim Crow, 55
life insurance and, 55–57
money lending and, 15–16
statistical analysis of differences and, 27–28, 226–27
Tuskegee study, 236–40
See also fairness/accountability/transparency concerns; privacy and justice; study of human differences
Raji, Inioluwa Deborah, 246
randomized controlled trials (RCTs), 86, 95
Rees, Mina, 96–97, 117, 203
Regan, Priscilla, 164
regression
causation and, 60
as foundational analytic tool, 66
Galton and, 39–40
Yule and, 58, 59–60, 61, 63
"Regression Towards Mediocrity in Hereditary Stature" (Galton), *40*
reification, ix, 31, 66–67, 70
"Road to Data Science, The" (Grus), 197–98
Roberts, Sarah, 219–20
Rockefeller Foundation, 119
Roeder, Ethan, 276
Rogaway, Phillip, 105
Rosenblatt, Frank, 177
Rosenthal, Caitlin, 15
Roth, Aaron, 253, 289, 295
Rudin, Cynthia, 289–90

SABRE (Semi-Automatic Business Research Environment) system, 142
SAGE (Semi-Automatic Ground Environment) system, 142
Salganik, Matthew, 235, 277
Schapire, Rob, 185
Schoolman, Carlota Fay, 256, 258
Schrage, Michael, 194
Schwartz, Jack, 138
scientific racism
eugenics and, 38–39
measure of distance and, 52–53
pseudoscientific physiognomy and machine learning and, 226–27
psychological application and, 71–73
science-based vs. political policymaking and, 58–59
science of individual differences and, 32–33
Tuskegee study and, 236–40
See also eugenics; fairness/accountability/transparency concerns; privacy and justice; racism; study of human differences
Seal, Brajendranath, 50–51, 53
self-regulatory organizations (SROs), 293–94
Selfridge, Oliver, 131
sequential analysis, 96
sequential decision theory, 181
Serra, Richard, 256, 258–59
Setler, Shivrang, 68
sexism
artificial intelligence debates and, 131–32
Bletchley Park and, 104
Galton and, 37–38
Google search and, 5–6, 17
invasions of privacy and, 153
statistical analysis of differences and, 27–28, 226–27
See also fairness/accountability/transparency concerns; privacy and justice
Shalizi, Cosma, 199
Shannon, Claude, 112, 120, 127–28, 133
Shmatikov, Vitaly, 249

Short, Jodi, 164
Simon, Herbert, 129–31, 133, 178, 183, 257, 258
Smith, C. R., 141
Smith, R. Blair, 141
Smith, Robert E., 156, *157*
Snowden, Edward, 6
Snyder, Samuel S., 115*n*
social change
data-driven internet and technology concerns and, 5–12, 306
data economic value and, 150–62
eugenics and, 50
future and, 306–7
mediation by data, x–xi, 4
Quetelet's social physics and, 18–20, 26–28, 29–32
The Belmont report, 233, 236, 238–43, 344*n*16
venture capital and, 279–83
See also privacy and justice; societal power relationships
social physics, 18–21, 26–28, 29–32
societal power relationships
historical understanding of, 18–23
quantification and classification of peoples and, 14–17, 38–39
transformative statistical analysis and, 56–57, 145
See also corporate power; people power; state power
Son, Masayoshi, 281
Soviet encryption system, 113
Spearman, Charles, 15, 68–70, *69*, 71–72, 171
Stapleton, Claire, 303
Stark, Luke, 227
state-driven capitalism, 144, 145
state power, xii, xiii, 305–7
adtech and persuasion architectures effects and, 276–78
algorithmic decision systems enabling of, 10, 158–60
corporate power defined and regulated by, 294–96
data-empowered technology distribution of, 8, 145–46
numerical accountability used for, 16–17
origins of statistics and, 22–25

state power (*continued*)
personal data collection and interpretation history and, 22, 143,
150–60
reconfigured antitrust regulation,
296–99
Section 230 reevaluation, 299–301
technological threats to individuals
and, 13–14
worldwide power and cryptography,
105–7
See also corporate power; people
power
Statistical Methods for Research Workers
(Fisher), 85
Statistical Society of London, 23
statistics, origins of, 21–34
applied computational statistics, 107,
112, 113, 116, 126, 140
author obscured data and, 82–83
Bayesian methods, 107–12, 116–17,
181
causation and, 49
criticisms of vulgar statistics, 20–21
hypothesis testing significance levels
and, 79, 81–83, 84–86, 89, 91–96
industrial-scale statistical analysis,
105, 107
innate differences and, 76, 226–27
original meaning, 21–22
policymaking and, 59–60
reform visions and, 18–19, 73, 75
statecraft, economies and, 22–25
transformative power, 56–57
Steingart, Alma, 124–25
Stigler, Stephen, 63, 109–10
Stop Online Piracy Act (SOPA) (2012),
293
"Strategic Subjects List," 10
"Student." See Gosset, William
Student's t-test, 81
study of human differences
data-driven racisms in, 50, 226–27
Galton and, 33, 37–38, 41, 71, 73
institutionalization of, 43
intelligence and race, 53
limitations of the data, 56, 57
Mahalanobis and, 52–53, 68
Pearson and, 71
Quetelet and, 27, 28

social hierarchies and, 76
social issues and, 58
Spearman and, 71
Yule and, 58
Stuetzle, Walter, 223
Suchman, Lucy, 123, 220
Super Pumped (Isaac), 292
support-vector machines (SVMs), 185
Suri, Siddharth, 219–20
surveillance capitalism
commercialization of the internet
and, 164–66
General Data Protection Regulation
(GDPR) and, 297
governments using business data for,
155–60
internet advertising microtargeting
and, 265–68
internet cloud-hosted computing
and, 224–25
NSA communications capability
expansion and, 6, 166–67
political persuasion architecture and,
274–78
prioritized economic efficiency over
human values, 160
privacy vs. free circulation of information, 152–54, 161–62
realignment of social order and
power inequities in, 6–10
See also NSA
Sutton, Rich, 185
Sweeney, Latanya, 249
systematic data profile analysis, 73–75,
145, 150–51

Taming of Chance, The (Hacking), 32–33
Taylor, Frederick, 15
technological determinism, 14, 306
Thatcher, Margaret, 31
theory of probability, 89
Tibshirani, Ryan, 226
Tilghman, Shirley, 199
Tooze, Adam, 75
Truth in Securities Act (1933), 15–16
Tufekci, Zeynep, 8, 255–56, 274, 275,
277
Tukey, John
on data analysis, 118–19
exploratory data analysis and, 201–5

on mathematical analytical judgment, 98
NSA and large data sets, 202, 206, 208
as renegade statistician, 222
Turing and, 112
Turing, Alan, 102, 103, 105, 111, 112, 121–23, 135
Turing Award, 114
Turing machine, 121
Turner, Fred, 165
Turow, Joseph, 265

unicorn, 261
United Kingdom
  GCHQ (Government Communications Headquarters), 6, 112, 166
  imperial decline, 35–36
  Poor Laws, 58, 65–66
United States v. Jones, 169
UNIVAC, 142, 146, 147, 169
U.S. Census, 22
U.S. Census Bureau, 96, 169, 170, 173, 250
U.S. Constitution, 22, 55
U.S. Department of Defense, 133–34
U.S. Department of Justice, 293
U.S. Immigration Act (1924), 71
U.S. National Security Agency (NSA). See NSA
U.S. presidential election (2016), 9
U.S. Public Health Service, 236–37
"U.S. Public Health Service Syphilis Study at Tuskegee," 236–40
user-generated content (UGC), 264–65

Vallor, Shannon, 246
Vapnik, Vladimir, 184, 187
VC: An American History (Nicholas), 280
venture capital, 279–83
Veysey, Victor, 152
Volinksy, Chris, 193

Wagner, Ben, 247–48
Wald, Abraham, 181
Wallach, Hanna, 3, 4, 250
Wanamaker, John, 269
Warner, Mark, 235
Weapons of Math Destruction (O'Neil), 6
Weber, James, 247
Wernimont, Jacqueline, 14, 23
Westin, Alan, 153
Wheeler, Stanton, 150
Whittaker, Meredith, 8, 191, 247, 303
Wickham, Hadley, 226
World War II, 96, 97, 98, 101, 105–6, 119, 279
World Wide Web, 211, 212–15, 260–64, 265
  See also internet
Worshipful Company of Drapers, 43, 44
Wu, Tim, 298

Yahoo, 212, 216
Yamamoto, Isoroku, 105–6
Yu, Bin, 221, 224
Yule, Udny, 58, 59–64, 65–66, 273

Zimmermann, Annette, 290–91, 304
Zuboff, Shoshana, 220, 255
Zuckerberg, Mark, 9, 160, 268, 269, 285–86